Textile Science and Clothing Technology

Series editor

Subramanian Senthilkannan Muthu, SGS Hong Kong Limited, Hong Kong, Hong Kong

More information about this series at http://www.springer.com/series/13111

Subramanian Senthilkannan Muthu
Editor

Textiles and Clothing Sustainability

Sustainable Textile Chemical Processes

 Springer

Editor
Subramanian Senthilkannan Muthu
SGS Hong Kong Limited
Hong Kong
Hong Kong

ISSN 2197-9863   ISSN 2197-9871   (electronic)
Textile Science and Clothing Technology
ISBN 978-981-10-9553-5   ISBN 978-981-10-2185-5   (eBook)
DOI 10.1007/978-981-10-2185-5

Printed on acid-free paper

This Springer imprint is published by Springer Nature
The registered company is Springer Science+Business Media Singapore Pte Ltd.

Contents

Ecological and Sustainable Natural Dyes

Nigar Merdan, Seyda Eyupoglu and Mujgan Nayci Duman

Abstract Since prehistoric times, natural dyes have been used to color of natural fibers such as cotton, wool, and silk as well as fur and leather. The use of natural dyes declined with the discovery of synthetic dyes in 1856. However, the increase in environmental consciousness created an upsurge in the interest in natural dyes. Nowadays, the use of natural dyes becomes common in food, cosmetic, pharmacology, and textile industry. In this study, natural dyes are investigated in all respects such as history, origin, chemical structure, advantages, and limitations. Furthermore, in natural dyeing, the innovative technologies are researched.

Keywords Natural dyes · Sustainability · Ecological · Innovative technologies

1 Introduction

The use of natural dyes trace back to ancient times. Through the ages, textile materials, leather, and foods have been dyed with natural dyes obtained from herbal and animal sources. Generally, natural dyes are provided from roots, steams, leaves, flowers, fruits of the plants, organism of the crustaceans, marine insects, snails, and cochineals. We want to explain that the synthetic dyes were discoverde and used because of natural dyes limitations. In textile industry, to synthetic dye stuff and pigments widely used because of their various colors, better color fastness properties, and low prices. However, the overlooked issue associated with synthetic dyes is that synthetic dyes are toxic, carcinogenic, and non-biodegradable. For these reasons, especially textile industry, the use of natural dyes is popular in our times.

N. Merdan · S. Eyupoglu (✉)
Engineering and Design Faculty, Department of Fashion and Textile Design,
Istanbul Commerce University, Kucukyali E5 Crossroad, 34840 Kucukyali, Istanbul, Turkey
e-mail: scanbolat@ticaret.edu.tr

M.N. Duman
Department of Textile Engineering, Institute of Sciences, Marmara University,
Istanbul, Turkey

© Springer Science+Business Media Singapore 2017   1
S.S. Muthu (ed.), *Textiles and Clothing Sustainability*,
Textile Science and Clothing Technology,
DOI 10.1007/978-981-10-2185-5_1

As far as the textile industry is concerned, dyeing and finishing process is an essential step so as to gain esthetic properties to textile materials. The dyeing of textile materials is carried out in an aqueous solution at high temperature. For this reason, the dyeing process of textile material demand high consumption of energy, water, dyes, and textile auxiliaries. Furthermore, the dyeing process causes environmental pollution so as to have dyes and textile auxiliaries [1]. Recently, the dyeing industry has been seeking for alternative process in order to reduce toxic effluents and environmental pollution. Therefore, the interest in natural dyes in textile industry has risen inasmuch as natural dyes are biodegradable, non-toxic, and eco-friendly [2–5].

In recent years, the increase use of synthetic dyes in textile, paper, paint, cosmetic, food, and pharmaceutical industries has caused to increase interest in the field of dye wastewater treatment for the global society. The degradable of synthetic dyes is very difficult and these dyes generally stable to light, oxidizing, and heat owing to their complex aromatic structure. Furthermore, another problem about synthetic dyes should be noted that these dyes have toxic, mutagenic, or carcinogenic compounds, they will have a huge impact on aquatic ecosystem and human health. Therefore, in the use of synthetic dyes, effective purification of wastewater becomes very essential. Nowadays, many techniques are used to purify of the wastewater such as membrane separation, ion exchange, flocculation, chemical oxidation, electrolysis, filtration, microbiological degradation, and photocatalysis degradation. Although these purification methods are low cost, high efficiency, and low energy consumption, the use of natural dyes eliminates the purification process of wastewater [6].

In this study, the use of natural dyes in textile dyeing process was investigated. Furthermore, chemical structure, origins, dyeing methods, advantages, disadvantages, toxicological, dermatologic, and antimicrobial effects and future of natural dyeing were researched.

2   Natural Dyes

The term "natural dyeing" has come to be used to refer to obtaining dyes from substances found in nature and to dyeing of various surfaces these dyes. The history of natural dyeing dates back to ancient times in the world.

3   History of Natural Dyes

In ancient times, humans used some substrates such as stones, soil, plants, and insects to dye textile materials. However, the interest in plants has increased in natural dyeing as the varieties of stones, soil, and insects are limited. Humans have begun, furthermore, to use flowers, leaves, and fruits of plants in dyeing acknowledging the variety of plant colors. As a result of archeological excavations, lots of substances have been used as natural dyes in order to dye textile materials.

- Indigo-dyed fabrics from the archeological finds, which go back to 3500 BC, were found to include indigo dye obtained from indigo plants of the genus Indigofera.
- In 3000 BC, madder (*Rubia tinctorium*) was used to dye purses produced from cotton fibers.
- The clay tablets, which date back to around 3000 BC in ancient Mesopotamia, were also found to include the dye derived from the kermes which yields the red color. Polish kermes and madder are, on the other hand, used in the samples of Pazyryk carpet, which are considered to be the oldest known carpet in the world and date back to 500 BC.
- The use of violet color obtained from sea snail has begun between the dates of 1800s BC and 1600s BC in the Mediterranean Seaside.
- In India, lac beetles were used to obtain red color in 1500 BCs.
- The use of cochineal beetles in dyeing dates back to 1000s BC in Mexico. Then, the dyeing with cochineal beetles has become widespread Europe and Asia.

The most important source of natural dyes is given in Fig. 1 and the chronological order of textile dyeing is given in Table 1.

4   The Classification of Natural Dyes

In the traditional system, natural dyes are graded on the basis of vegetable source and animal source.

4.1   Animal-Based Natural Dyes

In natural dyes, the number of colors of animal-based natural dyes is limited. The colors of this group are called "carmine red," "Indian yellow," and "violet." "Carmine red" is obtained from the beetles living in Cochenille plant. "Indian yellow" is obtained from camels and elephants living in India and China. Moreover, violet and red colors are obtained from sea snail and kermes insects, respectively [17–20]. Some animal-based natural dyes were given in Table 2.

Fig. 1 The most important natural dyes. **a** Indigofera plants and indigo dye [7, 8]. **b** Rubia tinctorum and madder [9, 10]. **c** Cochineal beetle and dye [11, 12]. **d** Lac beetle and dye [13]. **e** Sea snail and dye [14]. **f** Cinnebar and dye [15]

4.2 Plant-Based Natural Dyes

Plant-based natural dyes are obtained via extraction of plants' roots, barks, leaves, flowers, or seeds [22]. Some natural dyes along with their sources were given in Table 3.

4.3 Mineral-Based Natural Dyes

In ancient times, cinnabar, manganese oxide, and various copper salts used in wall paintings (Fig. 2) are the first mineral-based natural dyes. Ocher, which consists of hydrated alumina oxide, silica, and hydrated iron oxide, was used to dye nun clothes. Furthermore, ocher was used in wall painting with binder [42]. The mineral pigments from the ancient times are given in Table 4.

Table 1 The chronological order of textile dyeing [16]

2600 BC	The first written records regarding the use of natural dyes in China
715 BC	Handicraft of wool dyeing was established in Roma
331 BC	190-year-old violet cloths were found when Susa, the capital of Persia, was conquered by Iskender
327 BC	Alexander the Great mentioned the printing cotton fabrics in India
236 BC	Dyeing with papyrus plant was carried out in Egypt
55 BC	Romans dyed themselves with indigo dye
2 and 3 AD	Instead of the textiles dyed with violet purpur, madder, and indigo were found in Roman tombs. The papyrus that consists of the oldest receipt of dye was similarly found in a Roman tomb
The end of the fourth century	Byzantine Emperor Theodosius published the decree that the use of certain hue of violet was banned except for Imperial family
400	Natural dye obtained from Murex cost 20.000 $
700	Wax batik technique was mentioned in a China handwritten manuscripts
925	Wool dyeing was made commercial in Germany
1197	King John issued an order in wool dyeing in order to protect human health
1200s	The violet dye obtained from lichen was discovered in Florence
1212	In Ancient Florence, guidebook was prepared for weaver and thread chooser
1290	The blue dye obtained from woad became significant in Germany
1321	Natural dyes obtained from weld were discovered
1327–1377	In England, Edward III suggested to establish "Royal Wool Merchants Association" in order to develop the textile industry
At the beginning of the fifteenth century	Printing was achieved with natural dyes by using hand mold in Italy
1464	Pope Paul II introduced Cardinak violet obtained from Murex
1472	Edward IV established dyeing companies in London
1507	The cultivation of dye plants began in France, the Netherlands, and Germany
1519	Cotton was found in Central and South America, and the Indians used block printing techniques
1614	The textiles dyed with Brazil wood and fustics were introduced in England
1630	Drebbel, the Dutch chemist, produced a new bright red color by using red and tin, which was later used in England and Paris
1631–33	England began to import printed fabric from India
1688	James II banned the use of undyed textile materials in order to support dyeing industry
1689	The manufacture of batiste began in Germany

(continued)

Table 1 (continued)

Eighteenth century	English dyehouse patented obtaining dye from cochineal and application techniques
1708	William III banned the import of printed silk fabric in England
1716	In 1716, flax fibers were bleached with seaweed
1733	In England, John Kay discovered shuttle for weave machine
1745	After Industrial Revolution, the cultivation of indigo began in England
1766	Dr. Cuthbert Gordon patented the natural dye obtained from lichen
1774	Sulfuric acid and Prussian blue are the first commercial products in dyeing industry. Prussian blue was obtained with the use of potassium and iron salts
1786	Chlorinated compound and hydrogen peroxide began to be used as bleaching of fabric
1785	Roller printing technique was founded
1788	Picric acid was used as yellow color and disinfectant
1790	Discharge printing was developed
1794	Printing installation was established to produce printed batiste in France
1796	Tennat found the bleaching process
1797	Steam fixation of printing was developed
1802	Sir Robert Peel found reserve printing
1823	Mercer discovered the etching of indigo with chromate
1834	Runge, the German chemist, obtained the bright blue color by distillation of coal tar and addition of bleaching powder, which is the basis of aniline dye
1844	Mercer discovered the mercerizing process for cotton fibers
1856	William Henry Perkin discovered the first synthetic dye, aniline, when searching for a cure for malaria. Thus, synthetic dye has begun to industrialize

4.4 Microbial-Based and Fungal-Based Natural Dyes

In nature, some bacteria produce color materials based on secondary metabolites. *Bacillus Brevibacterium*, *Flavobacterium*, *Achromobacter*, *Pseudomonas*, and *Rhodococcus* spp. are some bacterias producing color pigments [43]. In literature, researchers investigated the natural dyes obtained from bacterias, microorganisms, and fungal because of economic properties of these dyes [44]. Vigneswaran et al. [44] investigated the dyeability of polyamide fabric using prodigiosin pigment extracted from *Serratia Marcescens*. In another study, *Monascus Purpureus* pigments obtained from Fungstan were used in the color of foods and fabrics. Further study showed that the washing fastness of silk and wool fabrics colored with *Trichoderma* sp. is very high [45, 46].

Table 2 Some animal-based natural dyes

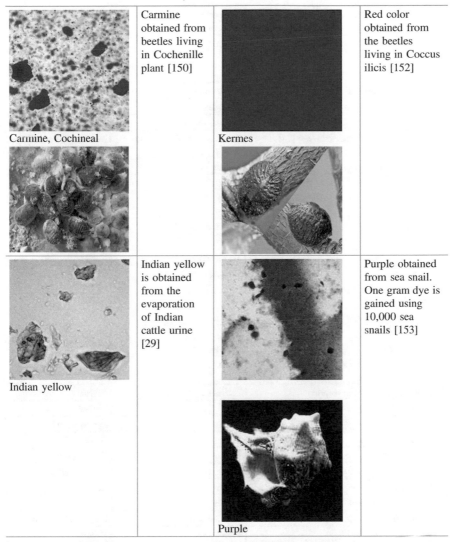

	Carmine obtained from beetles living in Cochenille plant [150]		Red color obtained from the beetles living in Coccus ilicis [152]
Carmine, Cochineal		Kermes	
Indian yellow	Indian yellow is obtained from the evaporation of Indian cattle urine [29]		Purple obtained from sea snail. One gram dye is gained using 10,000 sea snails [153]
		Purple	

Lichens and fungal were used as coloring source in all over the world. The purple dye called "orchil" obtained from lichens was used as an alternative to the expensive purple dye obtained from mollusks (Fig. 3).

The natural dyes obtained from mushroom have been popular since the 1970s. In particular, a considerable amount of dye is obtained from *Cortinarius* family which includes anthraquinone pigments such as emodin and dermocybin. Some mushroom types are given in Fig. 4. These pigments were investigated in a lot of studies in order to dye natural and synthetic fibers [48].

Table 3 The plants used in natural dyeing

The plants used as natural dye	The part from which the dye is obtained	Color
Salvia (*Salvia triloba*)	Trunk, leaves	Yellow
Phllyrea (*Phillyrea latifolia*)	Fruit	Green, gray
Grape leaves (*Vitis vinifera*)	Leaves	Yellow, green
Safflower (*Carthamus tinctorius*) [23]	Trunk, flowers	Yellow, green
Blueberry (*Vaccinium myrtillus*)	Fruit, leaves	Purple
Quince (*Cydonia vulgaris*)	Leaves, seeds	Yellow
Aesculus (*Aesculus hippocastanum*) [24]	Trunk coat, fruit coat and leaves	Mustard, milk coffee
Almond (*Prunus amygdalus stokes*)	Leaves	Yellow, green
Blackberry(*Rubus fructicosus*) [25]	Branch	Black, brown
Rhamnus (*Rhamnus tinctoria Rhamnus petiolaris*) [26]	Fruits	Yellow, mustard
Walnut (*Juglans regia* L.) [27]	Trunk, fruits	Green, brown
Isatis tinctoria [28]	Leaves	Blue
Blackthorn (*Prunus spinosa*)	Trunk coats	Camel
Carduus nutans	Handle, seeds	Straw yellow-straw brown
Morus alba	Fruits	Yellow
Hazelnut (*Corylus heterophylla*)	Leaves	Red
Alkanna tinctoria [29]	Root	Purple, green
Vitex agnus-castus	Leaves	Green
Elaeagnus (*Elseagnus oxycarpa*)	Leaves	Yellow, green-yellow, brown
Convallaria (*Convallaria majalis*)	Leaves	Yellow
Berberis (*Berberis crataegina*) [30]	Roots	Yellow
Prunus laurocerasus	Leaves	Lemon yellow
Pteridium aquilinum	Leaves	Green
Woadwaxen (*Genista tinctoria*)	Handle	Yellow-green
Astiibe (*Spirea hypericifolia*)	Trunk, leaves	Yellow, yellow-green
Thymus (*Thymus kotschyanus*)	Leaves	Beige, gray, khaki
Henna (*Lawsonia inermis*)	Leaves	Red, orange
Red onion (*Allium cepa*) [31, 32]	External leaf	Deep brown, orange
Alnus (*Alnus glutinosa*)	Trunk coats	Yellow
Pinus brutia	Trunk coats	Yellow, brown
Madder (*Rubia tinctorum*) [33]	Root	Orange, red, brown
Centaury (*Hypericum scabrum*) [34]	Leaves	Mustard, red brown, yellow-green
Wild cherry (*Prunus padus*)	Fruits, leaves, trunk coats	Yellow, green
Cistus (*Cistus creticus*)	Seeds	brown, black

(continued)

Table 3 (continued)

The plants used as natural dye	The part from which the dye is obtained	Color
Knotweed (*Polygonum cognatum*)	Leaves	Yellow, green, camel, gray
Quercus infectoria [35]	Fruits	Camel
Acorn Barnacle (*Quercus*)	Fruits	Cream, khaki, brown
Licorice (*Glycyrhiza Glabra*)	Root	Camel, green
Reseda luteola [36]	Every part of the vegetable	Green, yellow
Medlar (*Mespilus germanica*)	Leaves	Cinnamon
Myrtus communis L.	Leaves, fruits	Mustard, brown
Elderberry (*Sambucus nigra*) [37]	Fruits, leaves	Purple, mustard
Nane (*Mentha*)	Every part of the vegetable	Yellow, khaki, green
Nar (*Punica granatum*) [38]	Fruit coats	Yellow, brown, black
Mint (*Eucalyptus camaldulensis*)	Leaves, trunk coats	Green, brown
Valonia oak (*Quereus aegilops*) [39]	Fruits	Khaki, mustard
Beet	Trunk	Yellow
Daisy (*Anthemis tinctoria*)	Every part of the vegetable	Yellow
Saffron (*Crocus sativus*)	Flowers	Yellow
Resin (*Pistacia terebinthus*) menengiç	Leaves	Gray, purple
Gallum verum L.	Flowers, leaves, handle	Green, yellow
Mullein (*Verbascum phlomoides*) [40] *Verbascum phlomoides* L.	Every part of the vegetable	Mustard, green
Sumac (*Rhus coriaria*)	Every part of the vegetable	Yellow, red
Euphorbia (*Euphorbia* sp.)	Every part of the vegetable	Yellow, mustard
Rumex patientia	Root, seeds	Yellow
Pennyroyal (*Mentha pulegium* L.) [41]	Every part of the vegetable	Green, black

Fig. 2 The use of mineral-based dyes in wall painting [21]

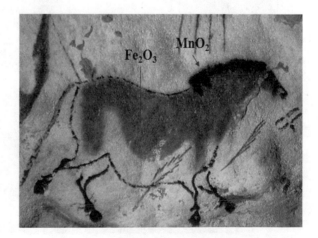

Table 4 Mineral pigments in ancient times [21]

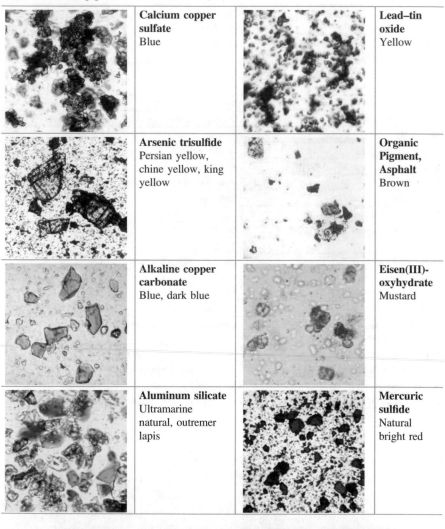

	Calcium copper sulfate Blue		**Lead–tin oxide** Yellow
	Arsenic trisulfide Persian yellow, chine yellow, king yellow		**Organic Pigment, Asphalt** Brown
	Alkaline copper carbonate Blue, dark blue		**Eisen(III)-oxyhydrate** Mustard
	Aluminum silicate Ultramarine natural, outremer lapis		**Mercuric sulfide** Natural bright red

Fig. 3 Some lichen species from which natural dyes are obtained [47]

Dictyophora cinnabarina   Cortinarius semisanguineus   Cortinarius violaceus

Fig. 4 Some mushroom types including color pigments [49–51]

5   The Chemical Structure of Natural Dyes

Natural dyes have complex chemical structures which can be classified into four main groups. These groups are as follows:

1. Indigoid dyes,
2. Quinone dyes,
3. Flavonoid dyes, and
4. Carotenoid dyes.

5.1   Indigoid Dyes

Indigoid dyes obtained from the indigo plant (*Indigofera tinctoria* L.), which is a tropic plant growing in India, and the color of these dyes is blue. The chemical structure of indigoid dyes is an aromatic compound derived from the combination of two indoxyl molecules, *Indigo suffraticosa* and *Isatis tinctoria* L., which have identical chemical structures. Furthermore, *Isatis tinctoria* L. is known as "*isatis tinctoria*" and is cultivated in Europe and Anatolia [52–54] (Fig. 5).

5.2   Quinone Dyes

Quinones are cyclic ketones that are derived from aromatic monocyclic and poly-cyclic components. Quinone groups produce the compounds that are of color

Fig. 5 6,6'dibromobenzene derivative of indigo

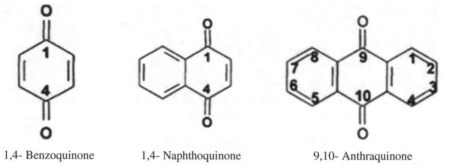

| 1,4- Benzoquinone | 1,4- Naphthoquinone | 9,10- Anthraquinone |

Fig. 6 The chemical structures of main quinone groups

quality. Quinones may be divided into three main categories such as benzo-quinones, naphthoquinones, and anthraquinones [19]. The chemical structures of these three main groups are given in Fig. 6 [19].

Natural dyes obtained from *Rhamnus* sp., *Fabaceae*, *Polgonaceae*, *Rhamnaceae*, *Dactylopuis Coccus Costa*, *Laccifer lacca Kerr*, and *Rubia tinctorum* L. are in quinone structure [55–57].

5.3 Flavonoid Dyes

In botanic, flavonoids, known as colorific groups, are compounds synthesized by plants in phenolic structure [58, 59]. In medical, flavonoids are utilized as anti-inflammatory, anti-allergic, anti-tumor formation, anti-viral, anti-diabetic, cardiovascular protective, antioxidant, and antimicrobial [58–61]. Flavonoids can be grouped into seven classes such as flavones, flavonols, isoflavones, antho-cyanidins, anthocyanins, and proanthocyanidins. The fundamental chemical structure of flavonoids is shown in Fig. 7.

Marčetić et al. investigated the natural dye extracted from venetian sumac (*Cotinus coggygria*). According to the results, the dye extracted from venetian sumac is in fisetin structure and the color of the dye which is of antioxidant and antimicrobial quality is yellow-brown [62].

Weld (*Reseda luteola*) and woadwaxen (*Genista tinctoria* L.) are in luteolin (Fig. 8) structure, anti-bacterial, and anti-inflammatory. The natural dye obtained from weld and woadwaxen is yellow.

Flavanones, the major group of flavonoids, are found in citrus. The chemical structure of flavanone is given in Fig. 9 [21].

In addition, isoflavones are found in pulses including soya (Fig. 10). Furthermore, apple, cacao, grape, bean, apricot, cherry, peach, green tea, black tea, and blackberry involve proanthocyanidin (Fig. 10) [63].

Fig. 7 The chemical
structure of flavonoid

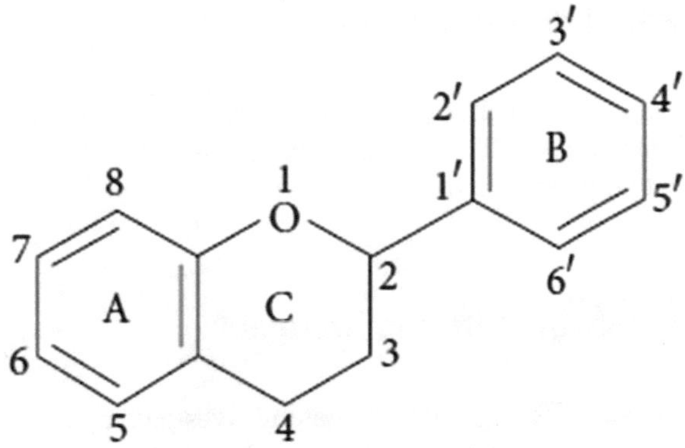

Fig. 8 Fisetin structure of
venetian sumac

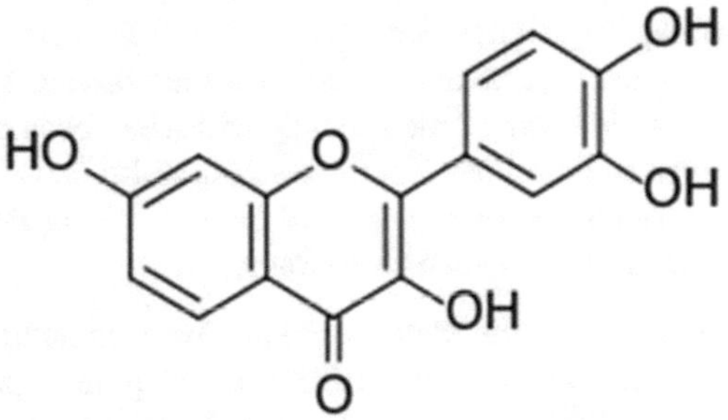

Fig. 9 Luteolin structure

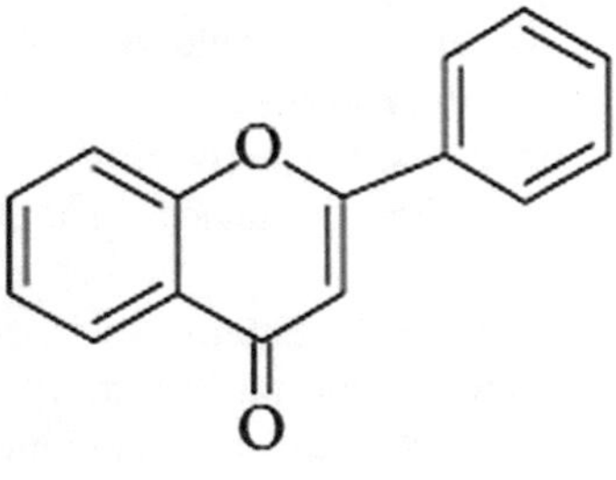

Fig. 10 Flavonone structure

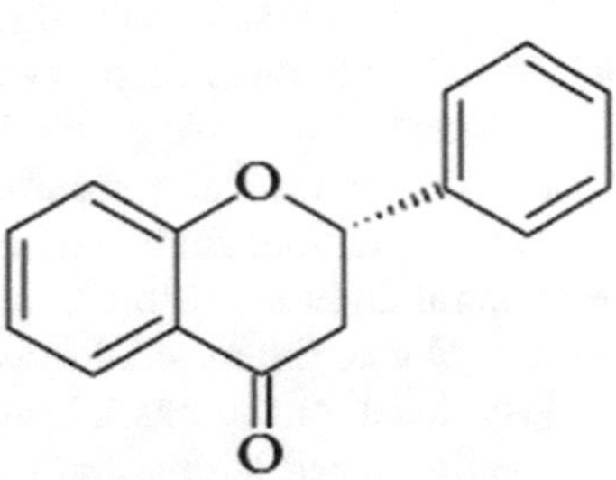

5.4 Carotenoid Dyes

Carotenoid chemical structure is found in some seeds of plants such as annatto. Furthermore, the flowers of parijat (*Nyctanthes arbor-tristis*) involve carotenoid chemical groups [54].

6 Natural Dyeing Methods

In textile dyeing industry, natural dyes are suitable to dye natural fibers. In recent years, natural dyes have been used to dye synthetic fibers [64]. Similar to synthetic dyes, natural dyes are used to dye fibers, yarns, and fabrics.

In natural dyeing, the dyeing process is manually achieved in iron, stainless steel, copper, and aluminum boilers. Besides, copper boilers are preferred for obtaining bright colors. Aluminum boilers can be used in uniform dyeing in order to keep stains. Stainless steel boilers are the most preferred boilers in dying processes. In fabric dyeing processes, jig and winchbeck machines are used. Moreover, natural dyes and auxiliary chemicals should be chosen depending on the desired color.

Though known as a very simple dyeing method, natural dyeing requires some special factors such as follows:

- Origin of the Materials to be Dyed: In natural dyeing processes, the origin of the material to be dyed becomes a quite significant issue in meeting the dyeing conditions. If a protein-based material is dyed, an acidic condition should be preferred in dyeing bath. On the contrary, if an alkaline condition is preferred, the material should be rinsed in a slightly acidic condition after the dyeing process and before drying. In addition, cellulose-based material should be dyed in an alkali condition. If cellulose-based material is, however, dyed in an acidic condition, the material should be rinsed in an alkali condition after the dyeing process.
- Mordant: In natural dyeing processes, choosing the type of mordant to be used is quite significant. The resulting color after dyeing is affected by numerous factors, including the mordant used. In mordanting process, the mordant bath should always be enough to let the fibers move around freely; water quality should be sufficient to dip the fabric/fiber properly.
- Temperature: Different natural dyes are more effective at different temperatures.
- Agitation: In order to get even dye uptake, it is necessary to move fibers around as much as possible in the bath. Wool fibers, however, should not be moved around too much as wool fibers tend to felt with high temperature and agitation.
- Bath Condition: Bath condition should be obtained according to the fibers.
- Natural Dyes are Unpredictable: In natural dyeing, the variety of factors affects the color of the material dyed. Some reasons for disappointing results could be insufficient or too much heat, accidental iron or other metal contamination, pH condition, and the harvest properties of plant.

- Rinsing: After the dyeing process, the textile material should be rinsed in a different condition in order to improve fastness properties. As mentioned above, the fibers should be returned to their optimum pH in the last rinse.

7 Extraction Methods of Natural Dyes

In nature, only a small percentage of materials consist of natural dyes. Apart from natural dyes, these materials include water-insoluble fibers, carbohydrates, proteins, chlorophyll, and tanen. In this regard, extraction is an important step in preparing and purifying natural dyes.

Natural coloring agents are not the only chemicals, including various non-plant components. Extraction of natural dyes is a complex process in which the nature of the coloring material and solubility characteristics should be determined prior to the process.

For centuries, natural dyes have been obtained via adding natural products directly to the dyeing bath, which is still common in India. Some disadvantages of the extraction method are, however, as follows:

- The separation of plant material from textile is difficult,
- The method is not proper for modern textile fabrication machines, and
- Natural dyes are not obtained from every plant with this extraction method.

In order to solve these problems of the traditional extraction method, innovative extraction methods have been currently used. The main innovative extraction methods are listed as follows:

- Aqueous extraction,
- Alkaline or acid extraction,
- Extraction with microwave or ultrasonic energy,
- Fermentation,
- Enzymatic extraction,
- Solvent extraction, and
- Supercritical fluid extraction.

7.1 Aqueous Extraction

Aqueous extraction is a conventional method to extract the plants and other materials. In order to enhance extraction effect, colored materials are initially cut, separated into little pieces or powdered, and sieved. Then, colored powders are stored overnight or even longer to loosen the cell structure in stainless steel containers. At the end of this process, the colored powders are boiled and dye solution is filtered. Boiling and filtering processes were repeated to obtain as much dye as possible.

When large amounts of dye extraction are carried out for the purification of powder materials stainless steel containers are used, and wetting time may be reduced by the boiling process. Centrifugation is typically used to remove residual material.

The trickling filter is used to provide a better resolution of the purified dye and to remove thin plant parts. Aqueous extraction is easily applicable to obtain natural dyes for textile materials. The disadvantage of this method is the long extraction time, high consumption of water, and high temperature. There are also other water-soluble materials used in extraction such as sugar. The extract may then be converted into powder form. Boiling in the heat-sensitive dyes decreases the yield. Extraction at a lower temperature is thus recommended.

7.2   Acid and Alkaline Extraction

Natural dyes generally contain glycosides and can be extracted in dilute acid and alkaline conditions. The addition of alkali or acid into the extraction bath accelerates the hydrolysis of glycosides while improving extraction. In addition, the color yield of extracted dye increases. Alkaline extraction is proper for natural products with phenolic groups [65]. After the extraction of natural dyes, the dyes can be precipitated with the use of acids. Alkaline extraction method can be used to extract lac dye from lac insect and safflower leaf. The disadvantages of this extraction method are that some natural dyes may be decayed in an alkaline condition. In addition, some natural dyes are pH sensitive.

7.3   Ultrasonic and Microwave Extraction

In these methods, ultrasound and microwave energy is used in order to enhance the impact of extraction process. The advantages of these methods are listed as low extraction temperature, saving chemicals and time. When ultrasonic energy is applied to a natural dye extraction, quite small air bubbles in liquid or cavity are created. After reaching a certain size, microbubbles retain their shapes. Here, millions of bubbles collide with each other every second, generating higher temperature and pressure while reducing the extraction time and increasing the extraction efficiency. Besides, better results are achieved with heat-sensitive dye molecules because of low temperature. Many researchers have used this method in order to optimize new dyes [66–69].

In microwave extraction, natural resources are subject to minimum processing and the amount of solvent is kept at minimum. It increases the speed of microwave processing, reducing the duration of processing, and yielding a better result. There are studies showing that dye is obtained from annatto plant and butterfly pea with this method. Saving energy, time, and chemicals, the methods of microwave and ultrasonic extraction are proved to be eco-friendly.

7.4 Fermentation

In this extraction method, enzymes produced by microorganisms are used (Fig. 11).

Firstly, indigo plant leaves are impregnated with water at 32 °C. In this method, colorless indigo occurs with indimusl enzyme which is in the leaves of indigo. Fermentation is completed in 10–15 h and yellow solution containing indoxyl beaten occurs after the process. After then, indoxyl is converted into blue indigotina insoluble and precipitates with the air (Fig. 12).

Indigo dyes can also be obtained from other plants containing indigo with fermentation process. Fermentation method is similar to the aqueous extraction method, but aqueous extraction method needs for high temperature. However, fermentation method has some disadvantages such as long duration of this process, the transition process immediately after harvest and bad smell due to microbial reactions.

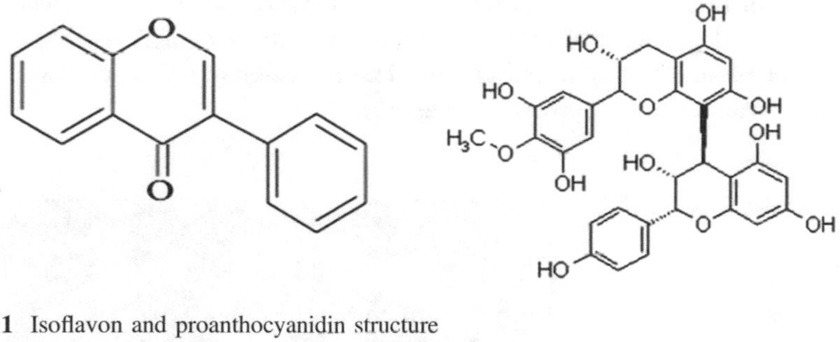

Fig. 11 Isoflavon and proanthocyanidin structure

Fig. 12 Preparation of indigo dyes with enzymatic extraction [70]

7.5   Enzymatic Extraction

In extraction of natural dyes, since plant tissue consists of cellulose, starches, and pectin, enzymes such as cellulose, amylase, and pectinase can be used in order to remove these molecules. Enzymatic extraction method can be used in mild conditions to improve the efficiency of extraction. This method may be generally suitable in extracting dye from hard plant materials such as bark and roots.

7.6   Solvent Extraction

Depending on the structure of natural coloring agents organic solvents such as acetone, petroleum ether, chloroform, ethanol, methanol, and mixture of alcohol and water mixture may be preferred (Fig. 2.20).

With water/alcohol extraction method, resources of both water-soluble and water-insoluble plant-based dyes are used. Compared to the aqueous method, the efficiency of extraction in water/alcohol extraction is higher.

Alcohol solvent, acid or alkali, is added to facilitate the collection of the hydrolysis of glycosides and color bodies. Purification of extracted dyes, removal of solvents by distillation as well as reuse, is easier. When processing at lower temperatures, less degradation exchange occurs. Toxic residues and greenhouse gas effect are the disadvantages of this method. The extraction of chlorophyll and waxy substances together creates problems (Fig. 13).

Fig. 13 Molded indigo dye [8]

Fig. 14 Solvent extraction [71]

7.7 Supercritical Fluid Extraction

Supercritical fluid extraction is a common technique in extraction and purification of natural products. The supercritical fluid is a function of gas above the critical temperature and pressure.

Supercritical fluid extraction method consists of two-stage process. The first step is using a dense gas as a solvent above its critical temperature. Having its critical temperature at 31 °C, carbon dioxide gas is ideal to dense gases and the. The second step is using a dense gas as a solvent above its critical pressure. After the determination of ideal gas in supercritical fluid extraction process, natural material is powdered and charged into the extractor. The ideal gas such as carbon dioxide is fed to the extractor with high pressure pump (100–350 bar). The extract charged carbon dioxide is sent to a separator with reduced temperature and pressure. After the reduction of temperature and pressure, the sediments of extract go out from the separator. In this method, carbon dioxide stream can be used several times for effective extraction of natural products [71]. Supercritical extractor is given in Fig. 14.

After the extraction process, natural dye extract should be filtered in order to remove macroparticles. The supercritical extraction method is defined as a clean, safe, inexpensive, non-toxic, environmental-friendly, and non-polluting process.

8 Mordanting

The first step of natural dyeing process is mordanting. A mordant is described as a metal salt to create an affinity between the fiber and the pigment. After applying mordant to a textile material, mordant attaches itself to the fiber molecules after which the dye molecule attaches itself to the mordant. In this way, mordanting process improves affinity for fiber and dye molecules as well as color fastness. Different colors derive from different mordant when combined with the same dye.

In general, mordanting process can be applied with three different methods. These methods are listed as follows;

Pre-mordanting (Onchrome): Firstly, the material is treated with mordant solution and then dyed with natural dye.

Meta-mordanting (Metachrome): The mordant and natural dyes are added together in the dyeing bath.

Post-mordanting (Afterchrome): Firstly, the material is dyed with natural dye after which the mordant is treated.

8.1 Metal Mordants

Metal mordants are materials that are commonly used with natural dyes in the textile dyeing. In the conventional dyeing process, metal salts are generally used such as aluminum, chromium, tin, copper, and iron salts. At the present time, because of the high waste load of chromium salts, these salts are evaluated as not-so-eco-friendly. For this reason, in natural dyeing, the use of chromium salts should not be preferred. Furthermore, copper is restricted in this category, but it can be used in very small amounts within the limits permitted.

Out of all types of alum, potassium aluminum sulfate is the most commonly used one in natural dyeing processes due to easy availability, cheapness, and safe use. Besides, potassium aluminum sulfate is an important mordant used in natural dyeing.

Another metal mordant used in natural dyeing is tin which gives brighter colors than other mordants. In textile natural dyeing, using tin too much is not recommended as it causes higher loss of fabric tensile strength.

Potassium dichromate (chrome), also referred to as red chromate, is known as being more expensive. In addition, being a heavy metal, potassium dichromate is harmful for human skin. The use of potassium dichromate has been limited as per the norms of the eco-standard.

In natural dyeing, copper can be applied very easily to textile materials. As being a heavy metal, however, the use of copper is limited under the eco-standard norms.

Tin is not limited to eco-label, yet it is not what is desired in terms of the waste load. Alum and iron can be considered as ecologically safe mordants. As a significant mordant used in natural dyeing, potassium alum sulfate is used at around 10–20 % of the material depending on color depth and dyestuff. This mordant cream can also be used with tartar as a mixture. In order for aluminum mordant to be properly fixated, the material is first processed with oils such as Turkish red oil or vegetable tannins/tannic acid and then mordanted with potassium alum sulfate.

There are many studies in the literature where metallic mordants are used in various dye resources and evaluated in terms of color strength dyeing fastness [71–74].

8.2   Mordanting Tannin and Tannic Acid and Oils

Tannin is a natural mordant agent that takes place in almost all parts of a plant, including roots, stems, bark, leaves, and fruits. Researchers deem that tannins emerge either in isolated individual cells in groups or in chains of cells. Tannins have complex molecular structures and high molecular sizes. Furthermore, initially classed with glycoside because of having glucose groups, tannins constituted a different class which did not have glucose groups later on. In addition, tannins involve C, H, O, N, P, and some inorganic elements in their structures [75].

9   Innovative Dyeing Methods

In 1960s, the use of natural products has increased with Flower-Power movement in the USA. Furthermore, leading to the consideration of most of the chemicals, a new concept called eco-textiles has emerged. Environmental movement for natural production against pollution has affected textile industry in 1990s and, consequently, the concept of textile ecology has emerged. Textile ecology involves ecology of textile production, human ecology, and residual ecology [76].

Due to global warming, decrease in water sources, chemical waste problems, and ecological legislations, ecological production methods became significant in all the industries. From a technical point of view, eco-friendly raw materials and eco-friendly production methods is of importance to environment [77].

Textile industries have been seeking for innovative and alternative technologies to meet both the needs of quality and eco-friendly production [78]. In textile dyeing and finishing industry, the use of plasma technology along with ultrasonic and microwave energy is the innovative and alternative technology to conventional methods. Plasma technology, newly treated to textile materials as a pre-treatment, causes a variety of chemical and physical modifications in textile materials. Plasma treatment also contributes to enhancing great numbers of textile properties such as hydrophobicity, dye exhaustion, and adhesion [79]. In addition, as a clean, eco-logic, and dry technique, plasma technology is characterized by low consumption of water, energy, and chemicals [80, 81]. In textile wet process, ultrasonic energy has been used as an alternative process to conventional method. With ultrasonic energy, textile wet process is performed at low temperatures, which, in turn, optimizes the use of heat energy. However, the materials dyed by conventional method give better fastness and color results than those dyed via ultrasonic method [82, 83]. Another technology is the use of microwave energy in textile wet process in order to conserve energy. In this technique, the liquor is heated using microwave energy, and microwave radiation provides more uniform heating in the liquor as opposed to the heating via conventional method. While the outer wall first heats up in conventional method, the center of the liquor initially heats up in microwave method. Subsequently, each point of solvent equally warms up. As such, the temperature of the liquor increases faster compared to conventional method. In this way, as an alternative method, microwave energy decreases energy consumption and saves time in comparison with the conventional method [84, 85].

9.1 Plasma Technology

Nowadays, the increase in the variety of productions to satisfy human needs has led to the abundant emergence of various industrial contaminations. In particular, water contamination and air pollution have increased day by day. In textile wet process, especially, water contamination is higher. Some techniques have been discovered in order to clean the wastewater. In this regard, conventional wastewater techniques comprised of biological, physical, and chemical processes are not effective in removing all pollutants from wastewater. In order to achieve removing all pollutants from wastewater, consumers have been seeking for eco-friendly innovation techniques among which plasma technology takes place and is used in textile wet process.

Plasma is an ionized gas that consists of electrons, neutrons, photons, free radicals, metastable excited species, as well as negative and positive ions. Furthermore, plasma is described as the fourth state of matter, which also provides environmental-friendly use in a wide range of industries. In 1879, the fourth state of matter was identified by Sir William Crookes for the first time. Moreover, Irving Langmuir coined the phrase "plasma" in 1929 [86].

Although gas state and plasma state are considered similar, there are also quite sharp differences between them. These differences are as follows:

- Gases do not conduct electricity, but plasmas are more conductive compared to metals.
- Unlike plasmas, gases are not affected by magnetic field.
- Compared to plasma state, reaction velocity of gases is low.
- While coulombic attraction of gases is limited between two particles, coulombic attraction of plasmas occurs across long distances.
- Plasma particles are not likely to fill spaces.
- Plasmas create electromagnetic field [87].

Plasma treatment can be classified as cold and hot plasma treatments. Compared to cold plasma, especially, hot plasma is not used in textile industry as it causes deformation of textile materials. Furthermore, hot plasma may lead to the carbonization of textile materials. For these reasons, in textile industry, cold plasma is used to achieve surface modifications. The types of plasmas are shown in Fig. 15.

Plasma technology is used in a wide array of industries owing to its numerous advantages, leading to the increase in the use of plasma technology in textile industry day by day. Table 5 shows the advantages of plasma technology compared to conventional methods in textile industry [78].

Plasma application led to surface modifications on textile materials such as etching, cleaning, grafting, cross-linking, and functionalization. On the surface of textile materials, weak covalent bonds are torn and then ions and radicals emerge with etching. Furthermore, the total surface area of textile materials increases and, thus, the adhesion properties of textile materials improve. Surface cleaning can be described as the process in which organic waste of the surfaces of textile materials is removed with ion bombardment, resulting in a sanitary textile surface. Grafting can be defined as the process of forming a thin polymer layer on textile surfaces. In plasma treatment, cross-links may occur on the surface of textile materials, leading to three-dimensional networks on the surface of textile materials. Consequently, the

Fig. 15 Supercritical extractor [88]

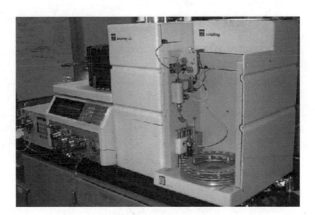

Table 5 The advantages of plasma treatment in accordance with conventional method

Plasma technology	Conventional methods
Water is not used in plasma application. Treatments are carried out in gas phase	Water-based
Consumption of water, energy, and chemicals is negligible	Consumption of water, energy and, chemicals is higher
Short period of application	Long period of application
Plasma application does not affect the bulk properties of materials	Bulk properties of materials are generally affected
Complex and multifunctional	Simpler
In plasma application, electric energy is used	In conventional method, heat energy is used

surface of material is more strength after the plasma treatment. After the plasma treatment, textile materials may gain some functional properties.

- Plasma treatment causes to gain anti-felting/shrink resistance of wool fabrics. Wool fabrics were treated with oxygen, nitrogen, and argon plasma under different conditions. After the plasma treatment, shrinkage of samples treated with oxygen, nitrogen, and argon plasma is 0 %. Furthermore, the results demonstrated that plasma treatment could impose significant anti-felting effects to the wool fabrics [89].
- Hydrophilicity of textile materials increases after the plasma treatment for improving wetting and dyeing. Hydrophilicity of cotton fabric treated atmospheric air plasma was investigated. According to the results, plasma treatment on cotton fabric improved the surface wettability. Moreover, the increasing of duration time of plasma treatment causes to increase in surface wettability [90].
- Hydrophilic enhancement for improving adhesive bonding. Noeske et al. investigated adhesion properties of polyethylene terepthalate, polyamide 6, polyvinylidene fluoride, polyethylene, and polypropylene treated with atmospheric plasma. According to the results, the adhesion of samples increased after the plasma treatment [91].
- After the plasma treatment, water- and oil-repellent properties of textile materials improve. Acrylic fabrics were processed with helium/oxygen plasma, and afterward, a fluorocarbon finish was applied through pad-dry-cure method. After plasma treatment, the water and oil repellencies of samples were measured. According to the results, the duration of water and oil repellencies of samples improved with the plasma treatment [92].
- Plasma treatment causes to facilitate the removal of sizing agents. Cai et al. investigated the effect of atmospheric plasma treatment on desizing of PVA on cotton fabrics. The results show that plasma treatment can remove some PVA sizing agent on cotton fabrics [93].
- Plasma treatment leads to scour of natural and synthetic fibers. Sun and Stylios investigated the effect of plasma treatment on scouring and dyeing properties of cotton and wool fabrics. The samples were treated with oxygen plasma at low frequency. According to the results, plasma treatment caused to improve scouring and dyeing processing by nearly 50 % [94].
- Anti-bacterial fabrics can be produced by deposition of silver particles in the presence of plasma. The possibility of using plasma treatment for loading of silver nanoparticles from colloids onto the polyester and polyamide fabrics and thus improve the anti-bacterial properties of fabrics was investigated. Plasma treatment caused to load of silver nanoparticles on the fabrics and the fabrics emerged anti-bacterial [95].

Furthermore, plasma treatment of fibers, yarns, and fabrics provides some functionalizations such as soil repellent, flame retardence, and electroconductivity [78].

9.2 Ultrasound Method

In textile industry, in order to meet the demands of consumers, producers have made innovation in textile wet process. In addition to innovation, saving water, chemical, and energy has been important to producers.

Ultrasound can be defined as a sound of which frequency is above the threshold of human hearing. A normal human can hear the sounds of which frequency is between 16 Hz and 16 kHz, whereas frequency of ultrasound is 20 kHz–500 MHz. For many years, ultrasonic energy is used in numerous fields such as engineering, medical, and science. Chart 1.1 shows in which areas and for which applications ultrasonic energy is used (Table 6).

Ultrasonic energy is processed with using transducer that converts electrical energy to sound energy. As a result of cavitation, chemical power of ultrasonic energy appears. The chemical power of ultrasonic energy derives from cavitation event. Cavitation is defined as the formation of microbubbles in the liquid on which negative pressure is sustained. Ultrasonic energy causes to occur cavitation in liquids and cavitation can accelerate chemical and physical reactions. Like a sound wave, ultrasonic energy is transmitted with sound waves. When the sufficient amount of a negative pressure is implemented on the liquid, fission is observed in the liquid and cavitation ballonets appear. In the successive compressing periods, these ballonets cause a huge among of energy to occur as they are colliding to each other [97].

Some examples of ultrasonic energy are military and non-military exploration of sea floor and sonar system, cleaning system of machine engineering, medical therapy, and diagnosis. In literature, there are many researchers that ultrasonic energy has been successfully applied to the textile wet process [98–101].

Ultrasonic energy causes to improve scouring, bleaching and dyeing, owing to cavitation, the phenomenon of expansion and collapse of microbubbles in the liquid. In the process of dyeing, cavitation can give rise to increase in four ways, (i) breaking up aggregates of dye in solution and diffusion rate of dye to fibers increase, (ii) removing air from between fibers to enhance contact of fiber–liquid,

Table 6 The using area of ultrasonic energy [96]

Field of application	Process
Biology and biochemistry	Decomposing of cell wall
Engineering	Treatment of grinded, cutting and perforating
Dental surgery	Carving and cleaning of tooth
Geology	Locating of minerals and petroleum stacks
Industry	Cleaning of engineering materials
Plastics and polymers	Welding process of thermoplastics polymers
Chemistry	– Observing of reaction rate – Quality control – Calculation of energy alteration – Measuring of liquid volume

(iii) infraction the fiber–liquid boundary layer to improve the diffusion rate of dye, and (iv) improving the swelling of fibers and increase in the area of the fiber–liquid interface [102], compared to the conventional process, the ultrasonic energy accelerates the chemical and physical reactions at the textile wet process [82, 103]. Ultrasonic energy effect results in the occurrence of textile wet process at low temperatures, which, in turn, optimizes the use of heat energy. Furthermore, compared to the fastness of materials which were dyed via conventional method, the color fastness of materials which were dyed via ultrasonic method is better [103, 104]. In literature, the ultrasonic method provides an eco-friendly dyeing process providing low dye bath temperature, shorter dyeing period, low electrolytes as well as low dyestuffs content [82, 105].

The advantage of using ultrasonic energy in textile wet process:

- In textile wet process, ultrasound energy improves the efficiency of diffusion rates of chemicals [106].
- Ultrasonic energy results in reduced dyeing time and consumption of dyeing auxiliaries [107].
- Compared to conventional dyeing in textile industry, ultrasonic energy increases color yield [108].
- Ultrasound has been proven to introduce several effects in solid–liquid systems such as the enchantment of mass transfer rate, the increase of the surface area by forming many microcracks on the solid surface and the clean-up of solid particle surfaces [109].
- Absorption/desorption rate significantly increases under sonication conditions [110].
- Ultrasound energy can be used in extraction process. Compared with traditional solvent extraction methods, ultrasound extraction can enhance efficiency of extraction, extraction rate, and extraction temperature [101].
- Ultrasonic energy gives rise to improve the effectiveness of subsequent oxidative–reductive bleaching in textile pre-treatment [111].
- Ultrasound technique has been also improved for the synthesis and deposition of nanoparticles on different textile substrates [112].
- Compared to conventional washing, ultrasonic energy has significantly effect on the strength and color of fibers during laundering while conserving the complete dirt removal with the same detergent concentration and laundering temperature [113].
- In laundry process, the use of ultrasonic energy results in less dimensional shrinkage, wrinkled appearance, and tensile rose [114].
- In textile scouring process, it can be told that the use of ultrasonic energy provides developments in washing efficiency. In other words, the use of ultrasonic energy in scouring leads to better whiteness degrees and removal of dirties [1].
- The use of ultrasonic energy results in good color strength and better washing fastness for dyeing of fibers. The development of fastness property is deemed to

the better covalent fixation of dye with the fibers, due to the better dye penetration into the fibers [115].

- Ultrasonic energy leads to improve the color yield of fabric [116].
- Application of ultrasonic energy for the extraction of high-cost materials has potential inasmuch as it is efficient, economical and simple extraction process compared to other extraction process [117].

The expansion of interest in ultrasonic energy in textile field has caused the improvement of range of applications from environmental protection. In textile wet process, ultrasound technology makes possibility to decrease the processing steps and to improve the product quality. Furthermore, ultrasonic energy brings many advantages to textile industry such as acceleration in process rate, increase in productivity, the level of product quality while reducing pollution, reduce temperature and reduce of using chemical, water, and energy. From all these reasons, ultrasonic energy has a promising future to textile industry [118–122].

9.3 Microwave Energy Technology

Microwave energy is a non-ionized electromagnetic radiation that has the frequency of 300–300,000 MHz. Spreading as electromagnetic radiations, microwave radiation can be used in a large number of areas. Microwaves were, indeed, originally used for communication technology. Doing some experiments about microwave radiation, Percy Spencer discovered the heating of materials with the use of microwave energy in 1946. After the use of microwave energy in food industry became popular in 1950s, it was also widely used in many industries such as chemistry, mechanics, and material industries. Microwave energy reduces the duration of process and save energy. For these reasons, microwave energy is used in many areas such as [123];

- Heating process,
- Drying process,
- Surface cleaning process,
- Melting process,
- Carbothermic reaction of oxide minerals,
- Synthesizing process [124],
- Galvanization process,
- Sintering process,
- Control bacterial contaminations [125],
- Waste sterilizations [125],
- Modification process [126],
- Calcination process [127],
- Heating metal matrix composites,
- Heating polymer matrix composites,

- Tempering process,
- Thawing process, and
- Curing of wood.

Microwaves are described as electromagnetic waves which contain electric and magnetic field with wavelengths in the range of 1–1000 mm. In case of interaction with metals, microwave energy is converted into heat energy. Since 1950s, microwave energy has been used in many applications. These applications include synthesis and drying applications, sintering applications, and advanced material processing applications. In this context, microwave process can be classified into three groups;

(a) Low-temperature processing: This process contains the use of microwave energy at the temperature below 500 °C. This process is used in food, wood, textile, and rubber industries.

(b) Moderate-temperature processing: The temperature of this process is between 500 and 1000 °C. This process is carried out in synthesis of carbon nanotubes, ceramic sintering, melting of glass, brazing, drilling of nonmetals, and heating of metals.

(c) High-temperature processing: This process involves utilization of microwaves at the temperatures above 1000 °C [128].

Microwave energy is commonly used all around the world in heating process, which is carried out at 915 and 2450 MHz in the household and industry [127, 129]. In conventional heating method, thermal energy is transferred from inside to outside. Unlike conventional heating methods, microwave drying method uses selective heating by targeting water molecules within the material directly, as electromagnetic field affects the material as a whole. According to this method, heat rises directly in the material. Moisture in the material is evaporated by heating in a short period of time and the transferring of moisture occurs from inside to outside due to the difference of vapor pressure inside and outside. As such, the problem of heat transfer in conventional heating methods goes away in microwave drying method [130].

In recent decades, the microwave energy has not only become an essential application in food industry but also has been used in textile wet process. Elshemy investigated the dyeing of wool fabric with cochineal extract by using microwave heating. The results showed that the microwave heating improved the fastness properties of samples [131]. Hou et al. investigated the effect of microwave irradiation on the physical properties and morphological structures of cotton cellulose. According to the results, the physical properties of cotton samples treated with microwave irradiation improved. Furthermore, the crystallinity and preferred orientation of the cotton samples treated with microwave irradiation increased [132]. Chun et al. dyed cotton fabric with reactive dyes via conventional and microwave heating methods. The results showed that the dye uptake and color fastness of samples via microwave heating are higher than the samples dyed via conventional method. Furthermore, microwave heating decreases the duration of dyeing process and saves energy [133]. In another study investigating the effect of microwave energy on the dyeing of acrylic fibers, it was found that the dyeability of acrylic fibers was significantly increased under microwave irritation owing to the local over heating, and accelerated the chemical reaction between fiber and dyes [134]. Haggag et al. investigated the effect of microwave radiation on the dyeability of polyester fabric with disperse dyes. It was found that microwave radiation accelerates the dye uptake and improves fastness properties of samples [135]. In another study, microwave heating has been applied to the process of dye-sensitized solar cells. The results showed that microwave heating dramatically improves the cell performances compared to conventional heating [136]. The effects of the use of microwave heating were investigated in terms of increasing reactive dye uptake and fixation on cotton fabrics. Compared to conventional heating methods, microwave heating yields better results with regard to improving color yield and dye fixation efficiency [137].

In textile wet process, microwave heating is more rapid, uniform, and efficient compared to conventional methods. Moreover, microwave energy provides easy penetration of particles into fibers, reducing the heat transfer problems.

10 Toxicological, Dermatologic, and Antimicrobial Effects of Natural Dyes

Natural dyes have been defined as using to color food substrate, leather, and natural fibers since pre-historic times. Since the advantages of synthetic dyes are found out such as availability, cheapness, excellent color fastness, and large color range, the use of natural dyes became poor in 1850s. However, the applications of natural dyes have recently increased with the increase in environmental awareness. The use of non-toxic, non-allergic, and eco-friendly natural dyes in textile industry has become a significant issue in order to avoid the hazardous effects of synthetic dyes. In spite of excellent performances of synthetic dyes, the use of natural dyes has been investigated in recent years by many researchers due to non-toxic and non-allergic properties. On the other hand, since the production of synthetic dyes depends on petrochemical sources, a variety of synthetic dyes consists of toxic and carcinogenic amines [138].

Toxicity can be defined as the ability of a substance to cause damage to a living tissue and nervous system and to result in a variety of illness when absorbed by living skin. In literature, LD_{50} is known as the figure corresponding to the toxicity rating of a substance. The term LD_{50} refers to the lethal dose for 50 % of the test animals. In nature, a variety of natural dyes are found as non-toxic, non-allergic, and non-carcinogenic [139].

In general, textile materials are known as being available to microbial attacks in order to absorb a substantial amount of moisture for microbial growth and to provide large surface area. In particular, natural fibers based on protein and cellulose provide oxygen, nutrients, temperature, and moisture for microbial growth.

Because of these properties of textile materials, researchers seek to develop antimicrobial and anti-bacterial agents for textile materials. Nanosized silver [140], basic magnesium hypochlorite [141], zinc oxide [142], silver nitrate [143], oxyte-tracycline chloride [144], oxolinic acid [144], flumequine [144], sarafloxacin [144], florfenicol [144], sulfadiazine [144], polymeric quaternary ammonium salts [145], heavy metals [146], halogens [146], fatty acid salts [147], and phosphonium salts [148] have antibacterial and antimicrobial quality. Antibacterial agents act against bacteria with five mechanisms, which are also inhibitions of cell metabolism of bacteria. These are inhibition of bacteria cell wall synthesis, interaction with the plasma membrane, disruption of protein synthesis and inhibition of nucleic acid transcription, and replication, respectively. Although a wide range of synthetic antibacterial agents show quite good toxicity against bacteria, using natural dyes and polymers with antibacterial quality can eliminate the environmental risk of synthetic antibacterial agents [149]. In recent years, in order to color and improve antibacterial quality of textile materials, the researchers have focused on the use of natural dyes. Turmeric [150, 151], pomegranate [150], Chinese gall [148], green tea [151], madder [151], saffron petals [151], henna [151], turmeric rhizomes [152], fenugreek seeds [152], oleander flower [152], weld [153], *Acacia catechu* [154], *Kerria lacca* [155], Quercus [155], *Rubia cordifolia* [155], *Rumex maritimus* [155], *Rubia tinctorum* [156], *Allium cepa* [156], *Punica granatum* L. [156], and Mentha sp. [156] are some natural dyes in order to dye textile material and improve antibacterial effect to textile materials.

11 Advantages of Natural Dyes

Natural dyes are derived from plants, minerals, and animal-based sources. Natural dyes were historically used to color clothes, textile materials, foods, etc. Founded out by the mid-1800s, the use of synthetic dyes gradually increased. Yet again natural dyes began to use as they are less toxic, less polluting, and environmental-friendly.

The advantages of natural dyes can be listed as follows:

* In order to derive from renewable sources, natural dyes are defined and widely accepted as eco-friendly across the world.
* Natural dyes are non-carcinogenic, non-allergic, and non-toxic for human skin.
* The fastness properties of natural dyes are acceptable with the use of proper mordants.
* After the natural dyeing process, the waste of process is biodegradable.
* Synthetic dyes are derived from non-renewable petroleum.
* Many natural dyes have significant protection to ultraviolet lights in order to absorb ultraviolet lights [157–163].

- Natural dyes have anti-bacterial, antifungal, and antimicrobial properties [150, 151, 163, 164, 165, 166, 167, 168, 169, 170, 171, 172, 173, 174, 175, 176, 177, 178, 179, 180, 181, 182, 183].
- Natural dyes have antioxidant agents in their structures, enhancing the antioxidant properties of materials [178–181].
- As some plants have insect-repellent properties, the natural dyes obtained from these plants have insect-repellent properties too.

12 Disadvantages of Natural Dyes

Synthetic dyes are used comprehensively in textile dyeing, paper printing, and leather dyeing, as well as additives in petroleum products. With the increasing use of the dyes in a variety of industries, the rate of pollution has become alarming. Although natural dyes have several advantages, there are some limitations as well.

The disadvantages of natural dyes can be listed as follows:

- The extraction process takes too much time, resulting in soil waste.
- Although textile dyeing with natural dyes is considered as environmental-friendly, there are many limitations too. One of the limitations of natural dyes is that dyeing process takes a longer period of time than synthetic dyes. Furthermore, in order to improve fastness properties, mordanting process can be treated.
- Purification of natural dyes for dyeing machine is not sufficient and costly.
- Sources of natural dyes are limited.
- After the dyeing process, a variety of dyes and mordant agents can emerge in the washing bath. Natural dyes are, however, biologically degradable.
- Color tones of natural dyes are limited.
- Natural dyes arc more expensive than synthetic dyes.
- Being one of the most important problems of natural dyes, repeatability is limited.
- Though requiring expensive and complex natural dyes, bright colors can be obtained in standard conditions.
- Some of the natural dyes are sensitive to pH changes, resulting in color changes occasionally.
- In natural dyeing, light fastness, washing fastness, and rubbing fastness are poor. In order to improve fastness properties of natural dyes, mordanting process is treated to textile materials before the dyeing [162, 184].
- Synthetic fibers are not dyed with natural dyes.
- Some mordants used in natural dyeing are not ecological.
- Some natural dyes have allergic effects to human [185, 186].
- Unlike synthetic dyes that are created in a laboratory, natural dyes are obtained from plants and are dependent on growing seasons.

13   Future of Natural Dyeing

Nowadays due to increasing environmental concerns around the world, synthetic dyes are losing its significance, replacing natural dyes. Widespread use of natural dyes is of major significance in terms of sustainability. Considering fossil resources, moreover, the wider use of renewable raw materials is important for the protection of the life cycle and farmland. In organic agriculture, the use of synthetic pesticides is not preferred. In view of environmental impacts, harvesting natural products with dye molecules is preferred. Furthermore, the use of plant waste as a resource of natural dye is quite important for recycling and sustainability. In textile dyeing process, therefore, natural dyeing is of significance with regard to sustainability and ecological approach (Figs. 16 and 17).

In the twenty-first century, natural dyes have not met economical and technical requirements such as repeatability. Despite the disadvantages of natural dyes, the increasing number of environmental concerns should be disregarded. Natural dyes do not have disadvantages of synthetic dyes such as health risk problem and high environmental pollution.

There is a consumer demand for "natural" products incorporating natural ingredients. Thus, investigation of natural products having natural dyes has become the focus of researchers. According to some researchers, nearly all natural dyes need application with a mordant to improve washing, rubbing, and light fastness. In literature, however, a variety of natural dyes obtained from natural plants have brilliant color, light, washing, and rubbing fastness properties.

In dyeing industry, there are some challenges of using natural dyes. To the extent that such challenges are responded, natural dyes can be used in industry. The major problem of natural dyes is that they are unable to meet customer expectations. Sustainability and eco-friendliness, however, are more important than ever.

In the last decade, the use of natural dyes in textile dyeing industry has been investigated by various research groups. In general, researchers precipitated to

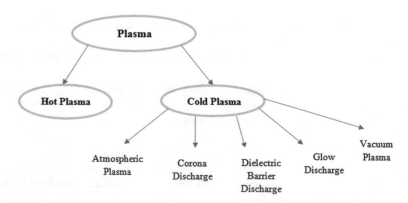

Fig. 16 Types of plasmas

Fig. 17 Lifecycle of natural dyes

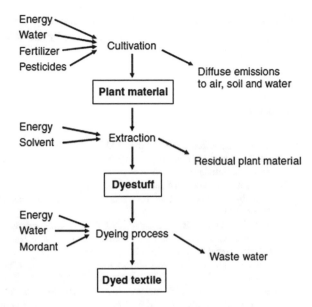

improve of fastness properties and color strength of natural dyes. Furthermore, according to the results, different mordanting and post-treatment can be used to increase color fastness properties of natural dyes [186]. As a result, various optimizations of natural dyeing process can be carried out.

Nowadays, there are numerous controversial discussions of the expected advantages of the use of natural dyes in future. While some researchers focus on the difficulties of dyeing processes of natural dyes, the others are interested in the sustainable sources of natural dyes [186]. The use of natural dyes in modern dyeing processes improves the sustainability of textile dyeing and finishing process with regard to water, chemicals, and energy consumption. As a result of the use of natural dyes in textile wet process, there is a decrease in the reduction of harmful chemicals in clothes.

Regarding the color range, the full range of colors might not be available and the quality of raw materials may not stable yet it is necessary to guarantee a defined reproducibility of colors. This can be achieved through various measures such as mixing dyestuff and varying the dye concentration of the dyeing bath.

In consideration of consumers, natural dyes are accepted by consumers because of non-carcinogenic, non-toxic, and non-allergic qualities. Although market research shows that natural dyes are believed to be less stable, they imply ecological awareness and responsibility. Moreover, it is known that natural dyes are likely to cause allergic reactions in some people. In order to reduce anxiety of consumers, non-toxic and non-allergic effects of natural dyes should be introduced to consumers.

In conclusion, professional marketing and economic research are important for natural textiles dyed with dyestuff from organic farming. In addition, there is a need for systematic information and optimization of current techniques. "The natural dyeing process" has yet to be developed and linked with the multitude of local and regional initiatives.

References

1. Tissera ND, Wijesena RN, Silva KMN (2016) Ultrasound energy to accelerate dye uptake and dye–fiber interaction of reactive dye on knitted cotton fabric at low temperatures. Ultrason Sonochem 29:270–278
2. Bechtold T, Amalid MA, Mussak R (2007) Natural dyes in modern textile dyehouses—how to combine experiences of two centuries to meet the demands of the future? Dyes Pigm 75:287–293
3. Haar S, Schrader E, Gatewood BM (2013) Comparison of aluminum mordants on the colorfastness of natural dyes on cotton. Clothing Text Res J 31(2):97–108
4. Barka N, Assabbane A, Nounah A, Laanab L, Ichou YL (2009) Removal textile dyes from aqueous solution by natural phosphate as new adsorbent. Desalination 235:264–275
5. Baliarsingh S, Panda AK, Jena J, Das T, Das NB (2012) Exploring sustainable technique on natural dye extraction from native plants for textile: identification of colourants, colourimetric analysis of dyed yarns and their antimicrobial evaluation. J Clea Prod 37:257–264
6. Li Y, Nie W, Chen P, Zhou Y (2016) Preparation and characterization of sulfonated poly (styrene-alt-maleic anhydride) and its selective removal of cationic dyes. Colloid Surf A: Physicochem Eng Aspects 499:46–53
7. http://medusasgarden.blogspot.com.tr/2012/07/true-indigo-dye-plant.html. 09 May 2016
8. https://www.pinterest.com/pin/264727284316589787/. 09 May 2016
9. http://wbd.etibioinformatics.nl/bis/flora.php?menuentry=soorten&id=3796. 09 May 2016
10. http://www.henriettes-herb.com/galleries/photos/r/ru/rubia-tinctorum-9.html. 09 May 2016
11. http://www.health.com/health/gallery/0,,20588763_15,00.html. 09 May 2016
12. http://www.bkherb.com/NaturalFoodColor/11/. 09 May 2016
13. https://middleagedjock.wordpress.com/2010/10/28/whats-in-it-wednesday%E2%80%94halloween-edition/. 09 May 2016
14. http://www.azerbaijanrugs.com/arfp-natural_dyes_dyestuffs.htm. 09 May 2016
15. http://www.naturalpigments.com/art-supply-education/vermilion-cinnabar-toxicology-test-results. 09 May 2016
16. Hofenk G, Roelofs WG (1999) The analysis of flavonoids in natural yellow dyestuffs occurring in ancient textiles, 12th Triennal Meeting
17. Rastogi D, Gulrajani ML, Gupta P (2000) Application of lac dye on cationised cotton. Colorage 47(4):36–40
18. Saxena S, Iyer V, Shaikh AI, Shenai VA (1997) Dyeing of cotton with lac dye. Colourage 44 (11):23–28
19. Chairat M, Rattanaphani S, Bremner JB, Rattanaphani V (2008) Adsorption kinetic study of lac dyeing on cotton. Dyes Pigm 76:435–439
20. Mahmud-Ali A, Binder CF, Bechtold T (2012) Aluminium based dye lakes from plant extracts for textile coloration. Dyes Pigm 94:533–540
21. http://www.emrath.de/pigmente.htm; November 2015
22. Wangatia LM, Tadesse K, Moyo S (2015) Mango bark mordant for dyeing cotton with natural dye: fully eco-friendly natural dyeing. Int J Text Sci 4(2):36–41
23. Bozkırlı DO (2007) Obtaining natural dye by supercritical carbondioxide extraction of safflower (*Carthamus tinctorius*) and its applicability, Thesis of Master of Science, Gazi University

24. Kahvecioğlu H (2003) The colors obtained from horse chestnut (*Aesculus hippocastanum* L.) and the fastness values of these colors on wool carpet yarns. Doctor of Philosophy Thesis. Ankara University
25. Yüksel S (2002) Effect of storage on the color of blackberry nectars. Ondokuz Mayıs Univesity, Turkey
26. Önal A (1988) Devolopment of dyeing methods and color properties of wool fibers dyeing with *Rhamnus tinctoria*. Thesis of Master of Science. Erciyes University, Turkey
27. Camcı N (2004) Extraction of dyestuff from cocoon of walnut (*Juglans regia* L): dyeing of wool, feathered-leather and cotton. Thesis of Master of Science. Gaziosmanpaşa University, Turkey
28. Kızıl S (2000) Investigations on determination of suitable sowing density and dyeing properties of some woad species (*Isatis tinctoria* L., Isatis constricta davis). Doctor of Philosophy Thesis. Ankara Universty
29. Öztav F (2009) The investigation of usage of the plant of alkanet (*Alkanna tinctoria*) as cellulosic and protein fiberic dyestuff. Doctor of Philosophy Thesis. Gaziosmanpaşa Universty
30. Gedikli F (2006) Investigation of wallnut (*Juglans regia*), black mulberry (*Morus nigra*), barberry (*Berberidis crataegina*), madder (*Rubia tinctorum*) and alder (*Alnus glutonisa*) as a protein dye in polyacrylamide gel electrophoresis. Thesis of Master of Science. Gaziosmanpaşa Universty
31. Seyfikli D (2009) Investigation of dyeing properties of mordanted with willow extract of wooden and fiber samples. Thesis of Master of Science. Gaziosmanpaşa Universty
32. Gümrükçü G (2003) Kırmızı soğan kabuğundan elde edilen antosiyanin ile yünlü kumaşların boyanması. Thesis of Master of Science. Yıldız Technical Universty
33. Öztürk M (2012) The analyzing of the morphologies of *Rubia peregrina* L. and *Rubia tinctorum* L. plants and their roots? dying features in comparison. Thesis of Master of Science. Uludağ Universty
34. Eray F (2006) The analysis of the fabrics dyeing with the St. John's wort (*Hypericum scarbrum* L.) plant extraction substance and its properties. Thesis of Master of Science. Gazi Universty
35. Güneş A (2010) Obtaining natural pigments from the gall oak (Quercus infectoria Olivier) shellac. Thesis of Master of Science. Marmara Universty
36. Cücen E (2009) Production of natural pigments from weld (*Reseda luteola* L.). Thesis of Master of Science. Marmara Universty
37. Atkinson MD, Atkinson E (2002) *Sambucus nigra* L. J Ecol 90(5):895–923
38. Çelikboyun P (2015) Determining antimicrobial activity of different solvent extracts and dyed fabric sample of *Punica granatum* and *Ruscus aculeatus*. Balıkesir Universty, Thesis of Master of Science
39. Yıldız A (1999) Extraction of dyestuff from volania oak (*quercus cerris*) and dyeing of wool, cotton and feathered leather. Thesis of Master of Science. Gaziosmanpaşa Universty
40. Sönmez HY (1992) Useablity of *Verbascum asperiloides* and *Arnebia densiflora* extracts in wool dyeing. Doctor of Philosophy Thesis. Cumhuriyet Universty
41. Şanlı HS (2007) Dycing with some natural dyes and determining fastnesses of silk textile products. Gazi Üniversitesi Endüstriyel Sanatlar Eğitim Fakültesi Dergisi 21:55–78
42. Agarwal OP, Tiwari R (1989) Mineral pigments of India. In: Compendium of the national convention of natural dyes. National Handloom Development Corporation, Lucknow, Jaipur
43. Joshi VK, Attri D, Bala A, Bhushan S (2003) Microbial pigments. Indian Journal Biotechnol 2:362–369
44. Vigneswaran N, Saxena S, Kathe AA, Gayal SG, Balasubramanya RH (2004) Bacterial pigments for eco-friendly textile dyeing, The Textile Institute 83rd world conference, 23–27 May. Shanghai, China, pp 765–768
45. Gupta C, Sharma D, Aggarwal S, Nagpal N (2013) Pigment production from Trichoderma spp. for dyeing of silk and wool. Int J Sci Nat 4(2):351–355

46. Venil CK, Zakaria ZA, Ahmed WA (2013) Bacterial pigments and their applications. Process Biochem 48:1065–1079
47. http://www.fungimag.com/summer-2014-articles/LR2%20V7I2%2066-69%20Dies.pdf
48. Raisanen R, Nousiainen P, Hynninen PH (2001) Emodin and dermocybin natural anthraquinones as a high temperature disperse dye for polyester and polyamide. Text Res J 71:922–1020
49. http://setasextremadura.blogspot.com.tr/2012/11/phallus-indusiatus-dictyophora-duplicata.html. 09 May 2016
50. http://www.funghiitaliani.it/?showtopic=31965. 09 May 2016
51. http://www.discoverwildlife.com/gallery/fabulous-fungi-photo-gallery-agorastos-papatsanis. 09 May 2016
52. Karadağ R (2007) Natural dyeing. TC Culture and Tourism Ministry, Ankara
53. Saxena S, Raja ASM (2014) Natural dyes: sources, chemistry, application and sustainability issues. Text Sci Clothing Technol. doi:10.1007/978-981-287-065-0_2
54. Ferreira E, Quye A, McNab H, Hulme A (2003) LC-ion trap MS and PDA–HPLC—complementary techniques in the analysis of different components of flavonoid dyes: the example of Persian berries (Rhamnus sp.). Dyes Hist Archaeol 18:13–18
55. Don RD, Rokayaa MB, Münzbergová Z (2012) Binu Timsinac, Krishna Ram Bhattarai, A review of its botany, ethnobotany, phytochemistry and pharmacology. J Ethnopharmacol 141:761–774
56. Angelini LG, Pistelli L, Belloni P, Bertoli A, Panconesi S (1997) Rubia tinctorum a source of natural dyes: agronomic evaluation, quantitative analysis of alizarin and industrial assays. Ind Crops Prod 6:303–311
57. Kumar S, Pandey AK (2013) Chemistry and biological activities of flavonoids: an overview. Hindawi Publ Corp Sci World J, 1–17
58. Kumar JK, Sinha AK (2004) Resurgence of natural colourants: a holistic view. Nat Prod Lett 18(1):59–84
59. Forgacs E, Cserhati T (2002) Thin-layer chromatography of natural pigments: new advances. J Liq Chromatogr Relat Technol 25(10–11):1521–1541
60. Deshpande R, Chaturvedi A Phytochemical screening and antibacterial potential of natural dye: Plumeria rubra (L.). Sci Res Report 4(1): 31–34
61. Marčetić M, Bozic D, Milenkovic M, Malesevic N, Radulovic S, Kovacevic N (2013) Antimicrobial, antioxidant and anti-inflammatory activity of young shoots of the smoke tree, Cotinus coggygria Scop. Phytother Res 27(11):1658–1663
62. Han KH, Okada TK, Seo JM, Kim SJ, Sasaki K, Shimada KI, Fukushima M (2015) Characterisation of anthocyanins and proanthocyanidins of adzuki bean extracts and their antioxidant activity. J Funct Foods 14:692–701
63. Hussain T, Tausif M, Ashraf M (2015) A review of progress in the dyeing of eco-friendly aliphatic polyesterbased polylactic acid fabrics. J Clean Prod 108:476–483
64. Mohanty BC, Chandranouli KV, Nayak ND (1984) Natural dyeing processes of india. Calico Museum of textiles, Ahmedabad
65. Liu WJ, Cui YZ, Zhang L, Ren SF (2009) Study on extracting natural plant dyestuff by enzyme-ultrasonic method and its dyeing ability. J Fiber Bioeng Info 2(1):25–30
66. Mishra PK, Singh P, Gupta KK, Tiwari H, Srivastava H (2012) Extraction of natural dye from Dahlia variabilis using ultrasound. Indian J Fiber Text Res 37(1):83–86
67. Pradeep KM, Pratibha S, Kamal KG, Harish T, Pradeep S (2012) Extraction of natural dye from Dahlia variabilis using ultrasound. Indian J Fiber Text Res 12:83–86
68. Rahman NAA, Tumin SM, Tajuddin R (2013) Optimization of ultrasonic extraction method of natural dyes from Xylocarpus Moluccensis. Int J Biosci Biochem Bioinfo 3(1):53–55
69. https://www.google.com.tr/#q=La+couleur+teintures%2C+pigments+et+luminophores; November 2015
70. http://nptel.ac.in/courses/116104046/9. pdf; November 2015
71. Vankar PS (2007) Handbook on natural dyes for industrial applications. National Institute of Industrial Research, Kanpur, India

72. Ghouila H, Meksi N, Haddar W, Mhenni MF, Jannet HB (2012) Extraction, identification and dyeing studies of Isosalipurposide, a natural chalcone dye from *Acacia cyanophylla* flowers on wool. Ind Crops Prod 35:31–36
73. Kiumarsi R, Abomahboub R, Rashedi SM, Parvinzadeh M (2009) *Achillea millefolium*, a new source of natural dye for wool dyeing. Color Coat 2:87–93
74. Jothi D (2008) Extraction of natural dyes from African Marigold flower (*Tagetes Ereecta* L) for textile coloration. AUTEX Res J 8(2):49–53
75. Şenol D, Kurtoğlu N (2004) General aspects of textile and ecology—textile chemicals that have carcinogen and allergic effects. KSU J Sci Eng 7(1):26
76. Verschuren J, Herzele PV, Clerck KD, Kiekens P (2005) Influence of fiber surface purity on wicking properties of needle-punched nonwoven after oxygen plasma treatment. Text Res J 75(5):437–441
77. Sivakumar V, Swaminathan G, Rao PG, Muralidharan C, Mandal AB, Ramasami T (2010) Use of ultrasound in leather processing industry: effect of sonication on substrate and substances—New insights. Ultrason Sonochem 17:1054–1059
78. Sahahidi S, Rashidi A, Ghoranneviss M, Anvari A, Wiener J (2010) Plasma effect on anti-felting properties of wool fabrics. Surf Coat Technol 205:349–354
79. Omerogulları Z, Kut D (2012) Application of low frequency oxygen plasma treatment to polyester fabric to reduce the amount of flame retardant agent. Text Res J 82(6):613–621
80. Bhat NV, Netravali AN, Gore AV, Sathianarayanan MP, Arolkar GA, Deshmukh RR (2011) Surface modification of cotton fabrics using plasma technology. Text Res J 81(10):1014–1026
81. Mason TJ, Lormier JP (1988) Applications and uses of ultrasound in chemistry. Ellis Horwood Limited
82. Khatri Z, Memonb MH, Khatri A, Tanwari A (2011) Cold pad-batch dyeing method for cotton fabric dyeing with reactive dyes using ultrasonic energy. Ultrason Sonochem 18:1301–1307
83. Büyükakıncı BY (2012) Usage of microwave energy in turkish textile production sector. Energy Procedia 14:424–431
84. Ahmed NSE, El-Shishtawy RME (2010) The use of new technologies in coloration of textiles fibers. J Mat Sci 45(5):1143–1153
85. Wang SY, Chen CT (2010) Effect of allyl isothiocyanate on antioxidant enzyme activities, flavonoids and post-harvest fruit quality of blueberries (*Vaccinium corymbosum* L., cv. Duke). Food Chem 122:1153–1158
86. Kut D (2011) Plazma Teknolojisi Ders Notları. Bursa, Turkey
87. Shishoo R (2007) Plasma technologies for textiles. Woodhead Publishing, Cambridge, UK
88. Feng XX, Zhang LL, Chen JY, Zhang JC (2007) New insights into solar UV-protective properties of natural dye. J Clean Prod 15:366–372
89. Pandiyaraj KN, Selvarajan V (2008) Non-thermal plasma treatment for hydrophilicity improvement of grey cotton fabrics. J Mater Process Technol 199(1–3):130–139
90. Noeske M, Degenhardt J, Strudthoff S, Lommatzsch U (2004) Plasma jet treatment of five polymers at atmospheric pressure: surface modifications and the relevance for adhesion. Int J Adhes Adhes 24(2):171–177
91. Ceria A, Hauser P (2010) Atmospheric plasma treatment to improve durability of a water and oil repellent finishing for acrylic fabrics. Surf Coat Technol 204(9–10):1535–1541
92. Cai Z, Qui Y, Zhang C, Hwang YJ, Mccord M (2003) Effect of atmospheric plasma treatment on desizing of PVA on cotton. Text Res J 73(8):670–674
93. Sun D, Stylios GK (2004) Effect of low temperature plasma treatment on the scouring and dyeing of natural fabrics. Text Res J 74(9):751–756
94. Radetic M, Ilic V, Vodnik V, Dimitrijevic S, Jovancic P, Saponjic Z, Nedeljkovic JM (2008) Antibacterial effect of silver nanoparticles deposited on corona-treated polyester and polyamide fabrics. Polym Adv Technol 19(12):1816–1821
95. Oner E (2002) Sonakimya Ders Notları. Istanbul, Turkey

96. Mason TJ, Lormier JP (1988) Sonochemistry: theory, applications and uses of ultrasound in chemistry. Ellis Horwood Limited, USA
97. Thakore KA, Smith CB (1990) Application of ultrasound to textile wet processing, American Dyestuff Reporter, pp 45–47
98. Smith B, Melntosh G, Shanping S (1988) Ultrasound—a novel accelerant, American Dyestuff Reporter, pp 15–18
99. Oner E, Baser I, Acar K (1995) Use of ultrasonic energy in reactive dyeing cellulose fabrics. JSDC 111:279–281
100. Giehl A, Schäfer K, Höcker H (1998) Ultrasonic in wool dyeing—ready for practical application. Int Text Bull 90–95
101. McNeil SJ, McCall RA (2011) Ultrasound for wool dyeing and finishing. Ultrason Sonochem 18:401–406
102. Akalin M, Merdan N, Kocak D, Usta I (2004) Effects of ultrasonic energy on the wash fastness of reactive dyes. Ultrasonics 42:1–9
103. Guesmi A, Ladhari N, Sakli F (2013) Ultrasonic preparation of cationic cotton and its application in ultrasonic natural dyeing. Ultrason Sonochem 20:571–579
104. Ismal OE, Yıldırım L (2012) Eco friendly approaches to textile design, 1th international fashion and textile design conference. Antalya, Turkey
105. Kamel MM, Helmy HM, Mashaly HM, Kafafy HH (2010) Ultrasonic assisted dyeing: dyeing of acrylic fabrics C.I. Astrazon Basic Red 5BL 200 %. Ultrason Sonochem 17:92–97
106. Hao L, Wang R, Liu J, Liu R (2012) Ultrasound-assisted adsorption of anionic nanoscale pigment on cationized cotton fabrics. Carbohydr Polym 90:1420–1427
107. Kan CW, Yuen CWM (2005) Use of ultrasound in textile wet processing. Textile Asia 36:47–52
108. Milenkovic DD, Dašic PV, Veljkovic' VB (2016). Ultrasound-assisted adsorption of copper (II) ions on hazelnut shell activated carbon. Ultrason Sonochem 16: 557–563
109. Li Z, Li X, Xi H, Hua B (2002) Effects of ultrasound on adsorption equilibrium of phenol on polymeric adsorption resin. Chem Eng J 86(3):375–379
110. Sun Y, Liu D, Chen J, Ye X, Yu D (2011) Effects of different factors of ultrasound treatment on the extraction yield of the all-trans-b-carotene from citrus peels. Ultrason Sonochem 18:243–249
111. Harifi T, Montazer M (2015) A review on textile sonoprocessing: a special focus on sonosynthesis of nanomaterials on textile substrates. Ultrason Sonochem 23:1–10
112. Hurren C, Cookson P, Wang X (2008) The effects of ultrasonic agitation in laundering on the properties of wool fabrics. Ultrason Sonochem 15:1069–1074
113. Ma M, You L, Chen L, Zhou W (2014) Effects of ultrasonic laundering on the properties of silk fabrics. Text Res J. doi:10.1177/0040517514537370
114. Bahtiyari MI, Duran K (2013) A study on the usability of ultrasound in scouring of raw wool. J Clean Prod 41:283–290
115. El-Shistway RM, Kamel MM, Hanna HL (2003) Ultrasonic-assisted dyeing: II. Naylon fibre structure and comparative dyeing rate with reactive dyes. Polymer Int 52(3):381–388
116. Kulkarni VM, Rathod VK (2014) Mapping of an ultrasonic bath for ultrasound assisted extraction of mangiferin from Mangifera indica leaves. Ultrason Sonochem 21:606–611
117. Yıldız K, Alp A (1999) Using of microwave in metallurgical processes. Metalurji TMMOB 24(125):1300–4824
118. Deveoğlu O, Karadağ R (2011) Natural dyestuffs: an overview. Marmara Univ Inst Pure Appl Sci 23(1): 21–32
119. Marinus A (2011) Towards a history of plasma-universe theory. Proceedings of NPA, pp 662–663. http://www.ph.surrey.ac.uk/newsite/ugrad_uploads/Bartlett2006May01161639.pdf
120. Vankar PS, Shanker R, Srivastava J (2007) Ultrasonic dyeing of cotton fabric with aqueous extract of Eclipta alba. Dyes Pigm 72:33–37
121. Ferrero F, Periolatto M (2012) Ultrasound for low temperature dyeing of wool with acid dye. Ultrason Sonochem 19:601–606

122. Larik SA, Khatri A, Ali S, Kim SH (2015) Batchwise dyeing of bamboo cellulose fabric with reactive dye using ultrasonic energy. Ultrasonics 24:178–183
123. Hou A, Wang X, Wu L (2008) Effect of microwave irradiation on the physical properties and morphological structures of cotton cellulose. Carbohydrate Poly 74(4):934–937
124. Ojha SC, Chankhamhaengdecha S, Singhakaew S, Ounjai P, Janvilisri T (2016) Inactivation of Clostridium difficile spores by microwave irradiation. Anaerobe 38:14–20
125. Li H, Lin B, Yang W, Zheng C, Hong Y, Gao Y, Liu T, Wu S (2016) Experimental study on the petrophysical variation of different rank coals with microwave treatment. Int J Coal Geol 154–155:82–91
126. Chen G, Li L, Tao C, Zuohua L, Chen N, Peng J (2016) Effects of microwave heating on microstructures and structure properties of the manganese ore. J Alloy Compd 657:515–518
127. Mishra RR, Sharma AK (2016) Microwave–material interaction phenomena: heating mechanisms, challenges and opportunities in material processing. Compos A 81:78–97
128. Chen BY, Chen D, Kang ZT, Zhang YZ (2015) Preparation and microwave absorption properties of Ni-Co nanoferrites. J Alloy Compd 618:222–226
129. Karaaslan S (2012) Microwave-related drying of fruits and vegetables. Süleyman Demirel University Agricultural Faculty Journal 7(2):123–129
130. Elshemy NS (2011) Unconventional natural dyeing using microwave heating with cochineal as natural dyes. Res J Text Apparel 15(4):26–36
131. Hou A, Wang X, Wu L (2008) Effect of microwave irradiation on the physical properties and morphological structures of cotton cellulose. Carbohydrate Poly 74(4):934–937
132. Chun WC, Liang H (2008) Microwave dyeing of cotton fabric, dyeing and finishing. http://en.cnki.com.cn/Article_en/CJFDTotal-YIRA200802007.htm
133. Nourmohammadian F, Gholami MD (2008) An investigation of the dyeability of acrylic fiber via microwave irradiation, progress in color. Colorants Coat 1:57–63
134. Haggag BK, Hanna HL, Youssef BM, El-Shimy NS (1995) Dyeing polyester with microwave heating using disperse dyestuffs, American dyestuff report. http://infohouse.p2ric.org/ref/02/01708.pdf
135. Maitani MM, Tsukushi Y, Hansen NDJ, Sato Y, Mochizuki D, Suzuki E, Wada Y (2016) Low-temperature annealing of mesoscopic TiO$_2$ films by interfacial microwave heating applied to efficiency improvement of dye-sensitized solar cells. Solar Energy Mat Solar Cells 147:198–202
136. Khatri A, Peerzada MH, MohsinM, White M (2015) A review on developments in dyeing cotton fabrics with reactive dyes for reducing effluent pollution. J Clean Prod 87: 50–57
137. Samanta AK, Agarwal P (2009) Application of natural dyes on textiles. Indian J Fibre Text Res 34:384–399
138. Samanta AK, Konar A Dyeing of textiles with natural dyes. http://cdn.intechopen.com/pdfs-wm/23051.pdf
139. Lee HJ, Yeo SY, Jeong SH (2003) Antibacterial effect of nanosized silver colloidal solution on textile fabrics. J Mat Sci 38(10):2199–2204
140. Fang M, Chen JH, Xu XL, Yang PH, Hilderbrand HF (2006) Antibacterial activities of inorganic agents on six bacteria associated with oral infections by two susceptibility tests. Int J Antimicrob Agents 27(6):513–517
141. Wiarachai O, Thongchul N, Kiatkamjornwong S, Hoven VP (2012) Surface-queternized chitosan particles as an alternative and effective organic antibacterial material. Colloids Surf B: Biointerfaces 92:121–129
142. Rai M, Yadav A, Gade A (2009) Silver nanoparticles as a new generation of antimicrobials. Biotechnol Adv 27(1):76–83
143. Hektoen H, Barge JA, Hormazabal V, Yndestad M (1995) Persistence of antibacterial agents in marine sediments. Aquaculture 133(3–4):175–184
144. Lu G, Wu D, Fu R (2007) Studies on the synthesis and antibacterial activities of polymeric quaternary ammonium salts from dimethylaminoethyl methacrylate. React Funct Polym 67 (4):355–366

145. Bakhshi H, Yeganeh H, Ataei SM, Shokrgozar MA, Yari A, Eslami SNS (2013) Synthesis and characterization of antibacterial polyurethane coatings from quaternary ammonium salts functionalized soybean oil based polyols. Mat Sci Eng: C 33(1): 153–164
146. Masuda M, Era M, Kawahara T, Kanyama T, Morita H (2015) Antibacterial effect of fatty acid salts on oral bacteria. Biocontrol Sci 20(3):209–213
147. Pugachev MV, Shtyrlin NV, Sysoeva LP, Nikitina EV, Abdullin TI, Iksanova AG, Ilaeva AA, Musin RZ, Berdnikov EA, Shtyrlin YG (2013) Synthesis and antibacterial activity of novel phosphonium salts on the basis of pyridoxine. Bioorg Med Chem 21 (14):4388–4395
148. Ghaheh FS, Mortazavi SM, Alihosseini F, Fassihi A, Nateri AS, Abedi D (2014) Assessment of antibacterial activity of wool fabrics dyed with natural dyes. J Clean Prod 72:139–145
149. Prabhu KH, Teli MD (2014) Eco-dyeing using *Tamarindus indica* L. seed coat tannin as a natural mordant for textiles with antibacterial activity. J Saudi Chem Soc 18(6):864–872
150. Zhang B, Wang L, Luo L, King MW (2014) Natural dye extracted from Chinese gall—the application of color and antibacterial activity to wool fabric. J Clean Prod 80:204–210
151. Selvam RM, Athinarayanan G, Nanthini AUR, Singh AJAR, Kalirajan K, Selvakumar PM (2015) Extraction of natural dyes from *Curcuma longa,* Trigonella foenum graecum and Nerium oleander, plants and their application in antimicrobial fabric. Indus Crops Prod 70:84–90
152. Ghoranneviss M, Shahidi S, Anvari A, Motaghi Z, Wienner J, Slamborova I (2011) Influence of plasma sputtering treatment on natural dyeing and antibacterial activity of wool fabrics. Prog Org Coat 70(4):388–393
153. Singh R, Jain A, Panwar S, Gupta D, Khare SK (2005) Antimicrobial activity of some natural dyes. Dyes Pigm 66(2):99–102
154. Negi PS, Jayaprakasha GK, Rao LJM, Sakariah KK (1999) Antibacterial activity of turmeric oil: a byproduct from curcumin manufacture. J Agri Food Chem 47(10):4297–4300
155. Calıs A, Celik GY, Katırcıoglu H (2009) Antimicrobial effect of natural dyes on some pathogenic bacteria. Afr J Biotechnol 8(2):291–293
156. Chattopadhyay SN, Pan NC, Roy AK, Saxena S, Khan BA (2013) Development of natural dyed jute fabric with improved color yield and UV protection characteristics. J Text Inst 104 (8):808–818
157. Katarzyna SP, Kowalinski J (2008) Light fastness properties and UV protection factor of naturally dyed linen, hemp and silk. Conference on flax and other bast PLANTS, Saskatoon, Canada. Accessed 21–23 July 2008
158. Sarkar AK (2004) An evaluation of UV protection imparted by cotton fabrics dyed with natural colorants. BMC Dermatol. doi:10.1186/1471-5945-4-15
159. Ibrahim NA, El-Gamal AR, Gouda M, Mahrous F (2010) A new approach for natural dyeing and functional finishing of cotton cellulose. Carbohydr Polym 82:1205–1211
160. Sun SS, Tang RC (2011) Adsorption and UV protection properties of the extract from honeysuckle onto wool. Indus Eng Chem Res 50:4217–4224
161. Ibrahim AN, El-Zairy WM, El-Zairy MR, Ghazal HA (2013) Enhancing the UV-protection and antibacterial properties of Polyamide-6 fabric by natural dyeing. Text Light Indl Sci Technol (TLIST) 2(1):36–41
162. Cristea D, Vilarem G (2006) Improving light fastness of natural dyes on cotton yarn. Dyes Pigm 70:238–245
163. Prabhu KH, Teli MD (2011) Eco-dyeing using *Tamarindus indica* L. seed coat tannin as a natural mordant for textiles with antibacterial activity. J Saudi Chem Soc. doi:10.1016/j.jscs. 2011.10.014
164. Velmurugan P, Chae JC, Lakshmanaperumalasamy P, Yun BS, Lee KJ, Oh BT (2009) Assessment of the dyeing properties of pigments from five fungi and anti-bacterial activity of dyed cotton fabric and leather. Color Technol 125:334–341
165. Sathianarayanan MP, Bhat NV, Kokate SS, Walunj VE (2010) Antibacterial finish for cotton fabric from herbal products. Indian J Fiber Text Res 35:50–58

166. Ghaheh FS, Mortazavi SM, Alihosseini F, Fassihi A, Nateri AS, Abedi D (2014) Assessment of antibacterial activity of wool fabrics dyed with natural dyes. J Clea Prod 72:139–145
167. Datta S, Uddin MA, Afreen KS, Akter S, Bandyopadhyay A (2013) Assessment of antimicrobial effectiveness of natural dyed fabrics. Bangladesh J Sci Ind Res 48(3):179
168. Gupta D, Khare SK, Laha A (2004) Antimicrobial properties of natural dyes against gram—negative bacteria. Color Technol 120(4):167–171
169. Singh R, Jain A, Panwar S, Gupta D, Khare SK (2005) Antimicrobial activity of some natural dyes. Dyes Pigm 66:99–102
170. Deans SG, Ritchie G (1987) Antibacterial properties of plant essential oils. Int J Food Microbiol 5(2):165–180
171. Yusuf M, Ahmad A, Shahid M, Khan MI, Khan SA, Manzoor N, Mohammad F (2012) Assessment of colorimetric, antibacterial and antifungal properties of woollen yarn dyed with the extract of the leaves of henna (*Lawsonia inermis*). J Clean Prod 27:42–50
172. Souza VB, Fujita A, Thomazini M, Silva ER, Lucon JF, Genovese MI, Favaro-Trindade CS (2014) Functional properties and stability of spray-dried pigments from Bordo grape (*Vitis labrusca*) winemaking pomace. Food Chem 164:380–386
173. Giri Dev VR, Venugopal J, Sutha S, Deepika G, Ramakrishna S (2009) Dyeing and antimicrobial characteristics of chitosan treated wool fabrics with henna dye. Carbohydr Polym 75:646–650
174. Kılınc M, Canbolat S, Merdan N, Dayioglu H, Akin F (2015) Investigation of the color, fastness and antimicrobial properties of wool fabrics dyed with the natural dye extracted from the cone of *Chamaecyparis lawsoniana*. Procedia—Social Behav Sci 195:2152–2159
175. Baliarsingh S, Panda AK, Jena J, Das T, Das NB (2012) Exploring sustainable technique on natural dye extraction from native plants for textile: identification of colourants, colourimetric analysis of dyed yarns and their antimicrobial evaluation. J Clean Prod 37:257–264
176. Han S, Yang Y (2005) Antimicrobial activity of wool fabric treated with curcumin. Dyes Pigm 64:157–161
177. Hashem M, Ibrahim NA, El-Sayed WA, El-Husseiny S, El Enany E (2009) Enhancing antimicrobial properties of dyed and finished cotton fabrics. Carbohydr Polym 78:502–510
178. Khan MI, Ahmad A, Khan SA, Yusuf M, Shahid M, Manzoor N (2011) Assessment of antimicrobial activity of Catechu and its dyed substrate. J Clean Prod 19:1385–1394
179. Khan SA, Ahmad A, Khan MI, Yusuf M, Shahid M, Manzoor N (2012) Antimicrobial activity of wool yarn dyed with *Rheum emodi* L. (*Indian Rhubarb*). Dyes and Pigments 95:206–214
180. Lawhavinit O, Kongkathip N, Kongkathip B (2010) Antimicrobial activity of curcuminoids from *Curcuma longa* L. on pathogenic bacteria of shrimp and chicken. Kasetsart J. (Nat Sci), 44: 364–371
181. Mariselvam R, Ranjitsingh AJA, Kalirajan K (2012) Anti-microbial activity of turmeric natural dye against different bacterial strains. J Appl Pharmacol Sci 2:210–212
182. Gerson H (1975) Fungi toxicity of 1, 4-napthoquinones to *Candida albicans* and Trichophyton mentagrophytes. Can J Microbiol 21:197–205
183. Samanta AK, Agarwar P (2009) Application of single and mixtures of red sandalwood and other natural dyes for dyeing of jüte fabric: studies on colour parameters/colour fastness and compatibility. J Text Inst 100(7): 565–587
184. Han N-R, Park JY, Jang JB, Jeong HJ, Kim HM (2014) A natural dye, Niram improves atopic dermatitis through down-regulation of TSLP. Environ Toxic Pharmacol 38:982–990
185. Anitha K, Prasad SN (2007) Developing multiple natural dyes from flower parts of Gulmohur. Curr Sci 92(12):1681–1682
186. Bechtold T, Turcanu A, Ganglberger E, Geissler S (2003) Natural dyes in modern textile dyehouses—how to combine experiences of two centuries to meet the demands of the future? J Clean Prod 11(5):499–509

Challenges in Sustainable Wet Processing of Textiles

Sujata Saxena, A.S.M. Raja and A. Arputharaj

Abstract Textile wet processing is an important step in textile production as it adds maximum value to the textiles by improving its aesthetics, comfort and functional properties. However, as the name indicates, a large amount of water is used as the medium which during the processing operations gets contaminated with unfixed dyes, chemicals and auxiliaries and is discharged in the end as effluent. The cocktail of chemicals present in this effluent makes it difficult to treat or biodegrade. It creates pollution problems and leads to further demand for good-quality water for processing. Natural textile fibres such as cotton need water and agrochemicals for growing, and petroleum resources are utilized for the manufacture of synthetic fibres. Energy is also needed at various steps of textile manufacturing process. Textile industry is thus water-, chemical- and energy-intensive industry and puts a lot of strain on global resources. Textiles now no longer just fulfil the basic human need of clothing; they have rather become a fashion statement. Rising income levels across the globe have led to manifold increase in world's textile production and consumption in recent years. Average per capita world fibre consumption in the year 2014 was 11.4 kg and estimated that it may increase further in forthcoming years. Present times are therefore very challenging for textile processing industry as the quantum of textiles to be processed has greatly increased and the environmental regulations are getting stricter. As the economies of the countries improve, they move out of the textile processing activities, which then shift to lesser developed nations. This has resulted in lack of serious efforts on improving the sustainability of textile wet processing and poorer adoption of the available methods. The present status of textile wet processing technologies, various developments to tackle the sustainability issues and future prospects would be discussed in this chapter.

Keywords Water quality · Energy · Chemical intensive · Agrochemicals · Cotton · Synthetic fibres · Sustainable wet processes

S. Saxena (✉) · A.S.M. Raja · A. Arputharaj
Chemical and Biochemical Processing Division, ICAR-Central Institute for Research on Cotton Technology, Adenwala Road, Matunga, Mumbai, India
e-mail: saxenasujata@rediffmail.com

© Springer Science+Business Media Singapore 2017   43
S.S. Muthu (ed.), *Textiles and Clothing Sustainability*,
Textile Science and Clothing Technology,
DOI 10.1007/978-981-10-2185-5_2

1 Introduction

Textiles today not only fulfil the basic human need for clothing but are also a fashion statement. The rising income levels have started fast fashion trend which has tremendously increased per capita consumption of textiles in past few decades. It is estimated that the global textile and garment industry including textile, clothing, footwear and luxury fashion is currently worth nearly $3000 trillion [1]. Contribution of the textile industry to the GDP and economy of many countries is substantial. However, it is a highly input-intensive industry. Textile fibres, the raw material used for making textiles whether natural or man-made, have their own environmental impact. These are then converted into yarns and fabric which are subjected to wet processing to impart aesthetic and functional values. Textile materials are treated with a number of dyes, finishing chemicals and auxiliaries in the process, and as the name suggests, water is used as carrier which gets discharged at the end of the process. This discharged water is highly polluting in nature as it contains a cocktail of unfixed residual chemicals and requires extensive treatment before it can be disposed. Increasing demand for textiles is increasing the quantum of wastewater from textile wet processing and is straining the scarce freshwater resources of the world. Textile industry in China, which accounts for nearly 54 % of the world's total textile production, discharges over 2.5 billion tons of wastewater every year [1]. The use of harmful, hazardous and persistent chemicals in wet processing is harmful for the ecosystem and human health. Today, most of the textile wet processing operations have shifted from the developed nations to the developing nations. As industries in these countries lack the resources to tackle pollution problems, it is leading to environmental damage. The 'Dirty Laundry' report published by Greenpeace International [2] highlighted the issue of discharge of toxic chemicals from textile wastewaters in Chinese textile processing units. A later study by the same organization conducted across remote mountain regions of 10 countries across three continents found the hazardous per- and polyfluorinated chemicals (PFCs) to be present in all snow and many water samples [3]. Such release of hazardous, persistent substances in aquatic environment threatens the sustainability of textile production. Sustainability has been defined as 'the ability to meet the needs of the present without compromising the ability of future generation to meet their needs and desires' (World Commission on Environment and Development). Besides water and chemicals, textile wet processing also consumes a lot of energy which is needed for heating the process water, for intermittent drying of textile material and for running the machines which results in the emission of greenhouse gases and increases the carbon footprints of the textiles. This chapter attempts to analyse the major sustainability challenges at various steps of textile wet processing such as consumption of water, energy, carbon footprints and usage and disposal of harmful chemicals and looks into the currently available solutions for sustainable processing. Technological advances such as the use of enzymes and biomaterials, recent developments in dyes, finishing chemicals and auxiliaries and the developments in processing machinery for source reduction and better sustainability have been explored. The present status of the futuristic waterless technologies such as supercritical carbon oxide and plasma technologies has been examined. Role of various eco-labels and standards in promoting sustainable chemical processing, the social and economic issues in adoption of sustainable chemical processing and responsibilities of various stakeholders in making the textile processing operations sustainable have been discussed in this chapter.

2 Textile Wet Processing Practices and Sustainability Issues

Textile material to be dyed and finished is first made water absorbent to make the fibres accessible to dyes and finishing chemicals. Size applied for weaving operations and natural impurities present on the fibres such as waxes, pectins and colouring matter in case of cotton and other natural cellulosics, wax and gum present on wool and silk, respectively, and also the oil, lubricants and waxes applied to the textile during spinning and knitting operations are all removed during the preparatory processing. The prepared absorbent textile is then dyed with a dye suitable for the fibre, and appropriate finishing chemicals as per the end use requirement are then applied to the textiles. Thus, preparatory processing, dyeing and finishing are the major textile wet processing operations.

Challenges in sustainability of textile wet processing operations mainly relate to the consumption of water energy, chemicals and discharge of unexhausted chemicals. Current practices followed in various wet processing operations and their sustainability aspects have been examined in the following sections.

2.1 Preparatory Processing

Desizing or removal of size is usually the first preparatory process. Size added to facilitate fabric weaving affects the fabric wettability and hence needs to be removed to facilitate proper wetting. Process employed for desizing depends upon the nature of size applied. Sizing materials are mainly of two types—natural sizing agents which include native and degraded starch and starch derivatives and cellulose derivatives such as CMC and protein sizes and synthetic sizes which include polyvinyl alcohol (PVA), polyacrylates and styrene–maleic acid copolymers.

Cotton is the most important natural fibre used in the textile industries worldwide and accounts for nearly 40 % share of the total global fibre consumption. Starch-based sizing agents are generally used for sizing cotton yarns as these are economical and provide satisfactory weaving performance. These sizing materials can be removed by hydrolysis with dilute acids or enzymes such as amylases. Hot caustic soda or detergent treatment can also remove these sizing materials but is comparatively less efficient. Though the water requirement of the process is relatively small, starch hydrolysis products greatly increase the biological oxygen demand (BOD) of the desizing effluent, and therefore, desizing process is the main contributor to BOD in cotton processing. In general, about 50 % of the water pollution is due to high BOD wastewater from desizing that renders it unusable [4]. Oxidative desizing processes such as cold pad-batch process based on hydrogen peroxide with or without the addition of persulphate and the oxidative pad-steam alkaline cracking process with hydrogen peroxide or persulphate are comparatively harsh on fibre, but they also simultaneously scour the fabric by removing the impurities. Synthetic sizing materials such as PVA though higher in cost can be removed by simple treatment with hot water and can be recovered from the bath by techniques like ultrafiltration which not only reduces the cost by facilitating reuse but also reduces the pollution.

Natural and added impurities present on textile materials such as waxes, oil, lubricants, gums and dirt, etc. removed by scouring to make it clean and hydrophilic. Cotton fibres contain many non-cellulosic substances such as wax, pectin, and proteins, which are removed during the scouring process. Conventionally, scouring of cotton is carried out at temperatures reaching up to 120 °C in a strong alkaline medium. Auxiliaries such as wetting agents, emulsifiers and sequestering agents are also added to the scouring bath to improve the process efficiency. The process can be carried out in kiers in a discontinuous manner or on a continuous range. Silk scouring or degumming is carried out with soap/surfactants under mild conditions to remove the gummy material known as sericin from the fibres which are made of fibroin protein. The process results in about 24–28 % weight loss and gives the silk a lustrous appearance and soft handle. Protease enzymes can also be used which remove sericin by hydrolysis. Wool which is also a protein fibre is scoured with surfactants and emulsifiers to remove greasy matter. Synthetic fibres both man-made and regenerated are free from the inherent impurities and comparatively need only a mild scouring with surfactants to remove dust, oil, lubricants, spin finishes, etc. added during yarn and fabric manufacturing process. As the impurities are emulsified and converted to soluble form, the scouring process effluent has high BOD, COD, TDS and alkalinity [5]. As the amount of sericin and wool grease in silk and wool, respectively, is substantial, these should be recovered from the processing water as it would otherwise result in very high BOD. Recovered materials can be further utilized for varied applications. Wool grease recovered from the scouring effluent has been used to obtain lanolin and wool wax alcohols.

Scoured textile material is hydrophilic, but it still contains natural or added colouring matter. These impurities are degraded in bleaching operations by oxidative or reductive agents to make the appearance of the textile pure white. Mostly, oxidative agents are employed for bleaching. Chlorine-based bleaching agents such as hypochlorites were widely used in the past, but their use has now been restricted due to the generation of adsorbable organohalides (AOXs) [6]. Hydrogen peroxide is considered an eco-friendly alternative to chlorine bleaching as its degradation products, water and oxygen are harmless. However, this process requires a large amount of water, high temperature, inorganic/organic stabilizers and neutralizing agents. Auxiliary chemicals which are used in the bleaching bath such as phosphate-based peroxide stabilizers increase the TOC and COD values of effluents. Upon neutralization of highly alkaline waste baths, large amounts of salts are produced. Apart from water and chemicals, huge energy is also required for the peroxide bleaching because this process is carried out at temperatures ranging from 90 to 120 °C. Large amounts of water and treatment with chemicals like thiosulphate are needed to remove residual hydrogen peroxide from fabrics which would otherwise cause problems in dyeing especially in reactive dyeing where it would accelerate dye hydrolysis. A more specific enzyme-based process targeting only the coloured substances would be advantageous. Preparatory processing, especially of natural fibres, consumes a lot of water and generates effluent with high BOD and COD values. Cotton preparatory processes including bleaching consume nearly 38 % of the total water consumption for cotton processing [7].

Alkyl phenol ethoxylates (APEOs) such as nonylphenol ethoxylates and octylphenol ethoxylates have been widely used in the textile industry as wetting agents and detergents. They are also used in the preparatory processes like scouring and bleaching. APEOs have hormone-disruptive properties and are found to be toxic to aquatic organisms. They are non-biodegradable and highly persistent and thereby cause problems in the wastewater treatment [8].

2.2 Dyeing and Printing

Colour is imparted to the textiles by dyeing and printing processes employing dyes and pigments. Differentiation between dyes and chemicals was historically based on their solubility. Colouring matters which were soluble or could be made soluble in water were known as dyes, and the insoluble ones were designated as pigments. An aqueous solution or dispersion of dyes is used for dyeing, and it imparts solid shade all over the textile material, whereas in printing, surface application of dye in the form of a pattern using a dye/pigment printing paste thickened using a sizing material is used. Dyes used for commercial textile applications are mostly synthetic in origin being derived from aromatic chemicals found in coal tar and have poor biodegradability.

Not all dye used for dyeing textiles gets fixed onto it. Dye fixation depends upon the application class and dye chemistry. Amount of unfixed dye remaining in the dyebath during dyeing of various textile fibres is presented in Table 1.

The unfixed dye results in highly coloured effluent which has a highly detrimental effect on aquatic ecosystems. Colour affects the photosynthetic capability of aquatic plants by reducing the light availability. Besides, dyes and their degradation products may be toxic, carcinogenic or mutagenic [10]. Azo dyes derived from certain carcinogenic or suspected to be carcinogenic aromatic amines have been banned by Germany and then other EU countries since 1990s. Majority of synthetic

Table 1 Percentage of unfixed dyes in dyeing of different fibrous materials [9]

Fibre	Dye type	Unfixed dye %
Wool and nylon	Acid dyes/reactive dyes for wool	7–20
	Premetallized dyes	2–7
	After chromes	1–2
Cotton and viscose	Azoic dyes	5–10
	Reactive dyes	20–50
	Direct dyes	5–20
	Pigment	1
	Vat dyes	5–20
	Sulphur dyes	30–40
Polyester	Disperse	8–20
Acrylic	Modified basic	2–3

dyes used are non-biodegradable. Therefore, these do not get degraded during the conventional effluent treatment process and are rather precipitated and separated as solid sludge which is dried and sent for landfills where it has the potential to contaminate land and groundwater. Large amount of Glauber's or common salt used for better exhaustion of dyes contributes to high TDS (total dissolved solids) content of the dyeing effluent. High TDS water is unfit for human, industrial or agricultural use. It cannot be reduced by conventional treatment, and costly processes like reverse osmosis (RO) or evaporation are needed to reduce it.

Commercial dyes are not pure compounds with the content of active ingredient ranging from 20 to 80 %. Some of them may also contain heavy metals in their structure. Copper and chromium are present in some metalized direct, reactive and metal complex dyes [11]. Metal salts are also used in wool dyeing with mordant dyes. Potassium dichromate often used for the oxidation of vat and sulphur dyes on cellulosic materials would result in the discharge of Cr^{6+} in the effluent which is a known carcinogen. All this can increase the heavy metal content of the textiles which is restricted by many eco-labels and standards. Also, the effluent containing such unfixed dyes and chemicals would require special treatment to bring down the heavy metal content within the permissible limits.

Many dyeing auxiliaries used during dyeing process are non-biodegradable and not recyclable and increase the BOD and COD loads of the effluent. Carriers used to facilitate dyeing of polyester (PET) under milder conditions are toxic and may affect the health of workers besides creating problems in effluent treatment. Decomposition products of sodium hydrosulphite (hydros) and sodium sulphide used as reducing agent in vat and sulphur dyeing increase the sulphide content of the effluent creating disposal problem as it has the potential to form sulphuric acid by bacterial oxidation.

Effluent volume is lesser in printing, as dye and pigments are applied in the form of thick printing paste, but there are concerns about the volatile solvents used in printing pastes which release VOC (volatile organic compounds) in the environment. Unfixed dyes and pigment, however, get discharged as effluent during the subsequent washing-off process, and cleaning of printing screens and other equipments also consumes water and organic solvents and releases the chemicals present in printing pastes to the environment. Many dispersing agents and auxiliaries added to the printing pastes may not be eco-friendly as they may contain formaldehyde from the use of formaldehyde-based fixer/binder and other harmful chemicals. The use of PVC and phthalates in plastisol printing pastes is another issue of concern as both of these are harmful.

2.3   Textile Finishing Operations

Hand and functional characteristics of textiles are modified by using various finishing chemicals. Formaldehyde has been widely used in the manufacturing of a large number of textile auxiliaries such as dye fixing agents, softeners, and cross-linking resins. Formaldehyde-based resins have been used to impart easy care and durable press properties to cellulosic textiles by the formation of cross-links, and these have also been used in many finishing formulations as cross-linking agents to improve durability. Such textiles during use may release formaldehyde which may cause eye irritation, skin rashes and allergic reaction, and it has also been classified under group 1 'carcinogenic to humans' since 2004 by the International Agency for Research on Cancer (IARC).

Durable water, oil and stain repellency in textiles has been achieved by polymeric finishes having perfluoroalkyl chains having eight or more fluorinated carbons [12]. These long-chain fluorinated polymers often contain residual raw materials and trace levels of long-chain perfluoroalkyl acids (PFAAs) as impurities. The residual raw materials and the product themselves may degrade in the environment to form long-chain PFAAs. On account of their widespread use and low biodegradability, long-chain PFAAs including perfluorooctanoic acid (PFOA) and perfluorooctane sulphonate (PFOS) having toxicological properties of concern have been detected globally in the environment, wildlife and humans [13].

Textiles are often treated with antimicrobial agents for health and hygiene purposes and to prevent odour formation due to bacterial action on perspiration. Among synthetic antimicrobial chemicals, usage of triclosan (2, 4, 4-trichloro-2-hydroxydiphenyl ether) which was very popular due to its efficacy is now being restricted as it gets degraded to 2, 8-dichlorodibenzo-p-dioxin by the action of sunlight which being a polychlorinated dioxin is toxic.

Many finishes used for imparting flame-retardant properties to textiles are toxic. Brominated flame retardants such as polybrominated diphenyl ethers (PBDEs) have been found to be very toxic to humans [14] but are still in use for textile finishing [15]. They are chemically similar to polychlorinated biphenyls (PCBs), which have been banned in many countries. Some flame retardants have toxicity due to the heavy metal content in their structure. Antimony oxide-based flame retardants are also used, though the amount of antimony on textiles has been restricted by eco-labels.

Textiles finished using harmful chemicals, catalysts and auxiliaries are not only harmful to the users of these textiles but also an occupational health hazard to the workers in processing units. The unfixed chemicals get discharged into effluent and pollute the ecosystem affecting a large number of people. Most of the finishing processes involve curing at high temperature for fixing the finish onto the textile substrate which requires energy which can contribute to the carbon footprints of textiles if it is derived from the burning of fossil fuels.

3   Environmental Impact of Textile Wet Processing

It transpires from the discussions in the previous section that there are many sustainability issues in the current wet processing operations in the textile industry. It is highly chemical-intensive, a number of eco-unfriendly and non-biodegradable products are being used in these operations, and a large amount of these chemicals remains unused at the end of operations which gets discharged along with the process water as effluent. The toxic non-biodegradable chemicals used in chemical processing are difficult to remove from wastewaters and may require tertiary and further treatments, and a failure to do so results in environmental contamination. As most of the textile wet processing operations are now located in developing or less developed countries where most of the processing is carried out by small-scale units, which lack the resources to adopt modern expensive technology for proper treatment of this effluent, it is being discharged into the environment without sufficient treatment to make it harmless. This has resulted in large-scale environmental damage in the area around these units. These contaminants can enter the food chain, and their increasing content over the food chain due to bioaccumulation may result in harmful levels of these contaminants for humans and animals. It was observed in a study conducted by Rajkumar and Nagan [16] in and around Tirupur, a major textile processing hub in the southern part of India, that untreated and partially treated effluent discharged in the river Noyyal by textile dyeing units for the past 20 years has accumulated in soil and water at many locations and it has affected the groundwater, surface water, soil, fish and the natural ecosystem in Tirupur and downstream area. Apart from the detrimental effect of the effluent discharge from processing units, burning of fossil fuels to provide energy for the various processing operations also results in the emission of greenhouse gases which adds to the global warming and climate change concerns. Detailed discussion of the environmental impact of textile wet processing has been made in the following subsections.

3.1 Impact on Water Resources

As the effluent coming out from textile processing units is discharged into the water bodies, these are the first to be affected by the textile wet processing operations. A large quantity of good-quality water is needed for these operations as water is used as a medium and solvent for dissolution of process chemicals as well as washing-off agent, besides generation of steam for heating the process bath. The actual amount of water used for processing depends upon the type of material being processed, nature of dyes and finishing agents and type of processing machinery employed; on an average, 50–100 L water (world average) is required for processing one kg of textile material [17]. Natural fibres usually consume more water for processing in scouring and bleaching comparison with synthetic fibres as natural impurities present on these fibres need water for their removal. General requirement of water for dyeing in different machineries is presented in Table 2.

Most of the water used along with unfixed dyes, chemicals and auxiliaries is discharged as effluent at the end of wet processing operations. Effluents from textile industries are complex mixtures of chemicals varying in quantity and quality. Both inorganic and organic pollutants are present in these effluents. These have high chemical oxygen demand (COD), BOD, TDS, pH, total suspended solid (TSSs) values and low dissolved oxygen (DO) value as well as strong colour [18]. The major concern with colour is its aesthetic character at the point of discharge as also the visibility of the receiving waters [19, 20]. Inorganic pollutants are mostly metallic salts, and basic and acidic compounds. These inorganic components undergo different chemical and biochemical interactions in the water bodies and deteriorate water quality. The organic pollutants are both biodegradable and non-biodegradable in nature. The biodegradable organic components degrade water quality as oxygen is consumed during their decomposition which depletes the DO affecting the aquatic animal forms. The non-biodegradable organic components persist in the system for a long time and pass into the food chain. As these substances move up the food chain, bioaccumulation takes place, affecting animals and humans.

Clean good-quality water is essential for humans as well other life forms. It is also needed for industrial activities and irrigation purposes. Inorganic and organic waste mixed with textile wastewaters leads to changes in both biological and chemical parameters of the receiving water bodies [21]. The scarcity of clean water and pollution of freshwater has led to a situation in which one-fifth of the urban dwellers in developing countries and three quarters of their rural dwelling

Table 2 Comparison of water usage (litre/kg of material) by different dyeing machines [9]

Nature of the equipment	Direct	Reactive	Vat	Disperse
Jigger	7.1	16.0	10.1	
Beck	18.4	37.9		22.5
Beam				19.5
Continuous			4.1	

population do not have access to reasonably safe water supplies [22]. The eco-logical and toxicological problem resulting from the discharge of effluent from textile industry into the streams and other inland water bodies is a major issue all over the world, especially in developing countries where textiles contribute sub-stantially to the national GDP. A study conducted by Kumar et al. [23] in Haridwar, India, found that discharge of textile effluents into the stream had considerable negative effects on the water quality and the water was not suitable for human consumption and domestic use. Toxic substances present in textile effluent also affect plants and vegetation, and the polluted water adversely affects agricultural production. It was observed in a study at Lagos, Nigeria, that watering with treated textile mill effluent mixed stream water reduced the photosynthetic pigments by 59.9 % and resulted in 41 % growth inhibition [24].

A study conducted in the Narayanganj district of Bangladesh concluded that the values of various water quality parameters including heavy metals like cadmium (Cd), copper (Cu), lead (Pb) and chromium (Cr) in water samples collected in the vicinity of effluent discharge area of a textile mill were much higher than the prescribed standards, and therefore, the water around the experimental area was unsuitable for recreation, drinking, domestic, irrigation and industrial purposes [25]. In Ethiopia also, discharge from textile factory was found to make the Blue Nile River water highly polluted. The values of most metrics followed pollution gradient and showed deteriorated water quality conditions [26].

3.2 Energy Consumption and Carbon Footprints

Globally, textile and clothing sector significantly contributes to carbon footprints and greenhouse gas emissions due to its lengthy and complicated supply chain [27]. To determine the carbon footprint of any product, it is essential to find out the level of 'embedded energy' within the product; this is all of the energy utilized at each stage of the production process [28]. The carbon footprint of a textile product depends on the nature of the raw material, i.e. fibre used and the method adopted for its processing. Though the energy requirement of chemical processing is only about 38 % of the total energy requirement of the textile production [29], energy use is very inefficient. Apart from a small amount of energy required for running textile machines, most of the energy required for chemical processing is thermal. Table 3 compares the energy requirement of different stages of chemical processing.

Table 3 Energy requirement for processing of 1 kg of textile material [30]

S. No.	Process	Energy (MJ/kg)
1	Scouring without drying	5–18
2	Bleaching and drying	8–33
3	Bleaching, dyeing and drying	10–35
4	Dyeing, finishing and drying	8–18
5	Finishing	6–12

As many textile processing operations are conducted at high temperature, energy is required for heating the bath and intermittent fabric drying operations. Reducing the water consumption for processing may reduce the energy consumption because heating of water and its removal will require less energy. Hence, dyeing in cabinet hank dyeing machine requires more energy than cheese dyeing process as amount of liquor to be heated up is more [31]. The use of low liquor ratio machines would therefore result in energy savings. Conservation of energy can be made through process and machine modification, proper chemical recipes and new technologies. Heat loss to ambient air from machines operating at high temperature can be reduced by proper insulation of the machines. Steam pipes supplying the thermal energy are often leaky, poorly maintained and not properly insulated leading to energy losses. Proper insulation provides resistance to convectional heat transfer with less steam and fuel consumption in heating contents up to the required temperature. In addition, insulation reduces the outer surface temperatures, which lessens the risk of burns. A well-insulated system also reduces heat loss to ambient workspaces, which can make the working environment in the factory more comfortable.

Thermal energy in textile mills is usually supplied by steam generated by burning of fossil fuels which results in the emission of greenhouse gases and contributes to the primary carbon footprints of textiles. These depend upon the type of fossil fuel used as emissions of carbon dioxide from natural gas are only around half of those produced from coal. The use of renewable energy therefore has great potential in reducing the carbon footprints of textile processing industries.

Indirect carbon footprints of textiles are derived from the embedded energy of chemicals and raw materials used. Textile fibres differ widely in their carbon footprints with synthetic fibres having larger carbon footprints due to embedded energy costs in their manufacturing. But there is not much difference in the carbon footprints of these fibres at processing stage. Regarding the emission contribution of various processing steps, in a fully continuous textile finishing process for cotton textiles, about 50 % emissions come from drying, 40 % from washing and steaming and 10 % from the use of chemicals. In finishing of knitwear by exhaust process, 60 % emissions are caused by heating the water [32]. Apparel and textiles account for approximately 10 % of the total carbon impact in the world. The estimated consumption of electricity for the annual global production of 60 billion kilograms of fabrics was estimated to be 1 trillion kilowatt hours [33].

3.3 Use of Toxic and Non-biodegradable Products

About 8000 different chemicals are used in textile processing, and many of these as discussed in the earlier sections are toxicologically and environmentally harmful. It is estimated that about six million tonnes of textile chemicals are used every year [34]. The use of alkalis for scouring, hypochlorite for bleaching, APEO as wetting agents, formaldehyde-based auxiliaries and cross-linking agents, petroleum-derived non-biodegradable and heavy metal containing dyes, Glauber's and common salt for dyeing of cellulosics, carriers for dyeing of PET under milder conditions and use of bromine/antimony-based chemicals for functional finishing are some of the examples. The presence of these unfixed substances in the process wastewaters at the end of the process is an environmental hazard as its treatment and disposal is a problem due to the potential to pollute the land and water bodies wherever it is discharged. Besides, the use of toxic chemicals also poses an occupational health hazard to workers in the processing unit. It is therefore suggested that hazardous and non-eco-friendly chemicals used in wet processing should be substituted by less toxic and more eco-friendly products so that toxicity of the discharged effluent gets reduced and it has better biodegradability. Some such alternative chemicals for various processes to reduce effluent toxicity are listed in Table 4.

Table 4 Eco-friendly alternative chemicals for textile wet processing [35]

S. No.	Purpose	Chemical	Alternative
1	Sizing	Starch	Water-soluble polyvinyl alcohol
2	Desizing	Hydrochloric acid	Amylases
3	Scouring of cotton	Sodium hydroxide	Pectinases
4	Bleaching	Hypochlorites	Hydrogen peroxide
5	Oxidation of vat and sulphur dyes	Potassium dichromate	Hydrogen peroxide, sodium perborate
6	Thickener	Kerosene	Water-based polyacrylate copolymers
7	Hydrotropic agent	Urea	Dicyanamide (partially)
8	Water repellents	C8 fluorocarbons	C6 fluorocarbons
9	Crease recovery chemicals	Formaldehyde-based resin	Polycarboxylic acids
10	Wetting agents and detergents	Alkyl phenol ethoxylates	Fatty alcohol phenol ethoxylates
11	Neutralization agent	Acetic acid	Formic acid
12	Peroxide killer	Sodium thiosulphate	catalases
13	Mercerization	Sodium hydroxide	Liquid ammonia
14	Reducing agents	Sodium sulphide	Glucose, acetyl acetone, thiourea dioxide
15	Dyeing	Powder form of sulphur dyes	Prereduced dyes
16	Flame retardant	Bromated diphenyl ethers	Combination of inorganic salts and phosphonates
17	Shrink proofing	Chlorination	Plasma treatment

4 Technological Developments in Source Reduction for Mitigating Environmental Impact

As treatment of textile effluent to make it safe for disposal in a manner which is not harmful to the environment is difficult and costly, various attempts have been made to reduce the input requirement in textile processing which can reduce the pollutant load as also the quantity of effluent generated and thereby improve the sustainability of textile industry. Some of these approaches have been described in the following sections.

4.1 Use of Biomaterials and Renewable Energy

The use of materials of biological origin is an approach which has excellent potential to reduce the pollution load as these materials are biodegradable and are generally safe. Bioprocessing or the use of enzymes for textile preparatory and finishing processes, the use of natural dyes for coloration and the use of natural product-based materials for functional finishing of textiles are the fields where biomaterials have been used with varying degree of success and the same has been discussed in the following lines.

Bioprocessing Using Enzymes

Processing using enzymes generally requires less energy, less water and less effluent problems as enzymes being biocatalysts are readily biodegradable. Alpha amylase enzymes which degrade starches into glucose have successfully replaced the acidic desizing of starch-based sizes used for cellulosics. As enzymes are very specific in action, no fibre damage takes place during the desizing process, whereas acids have a degrading action on cellulose which also increases the effluent load besides the acidic nature of the effluent. The use of amylase for desizing can also reduce the consumption of water, chemicals and energy [36]. Both amylases and acids, however, degrade the starch to glucose which results in high BOD of effluent. Glucose oxidase enzyme has been used to convert the glucose generated during desizing into hydrogen peroxide [37]. Alkaline pectinase enzyme-based process can be an alternative to alkaline scouring at high temperature and pressure in kiers. The enzymatic scouring can be carried out at moderate temperature resulting in energy savings, and also the gentle action of the enzyme on cotton fabric results in less damage, less pilling and uniform dye uptake [38].

Huge savings in energy and water by bioscouring have also been illustrated by Menezes and Choudhari [39]. Enzymatic desizing cum scouring results in lower, about 2–3 % fibre loss as compared to 7–8 % in conventional process resulting in about 5 % lower TSS value. Hebeish et al. [40] have tried to combine enzyme-based scouring with activator-assisted bleaching in single bath to remove the non-cellulosic impurities from the cotton textiles at moderate temperature.

The use of laccases from different sources for removal of colouring substances from cotton textiles has recently been looked into to make the process more eco-friendly. Though hydrogen peroxide is considered an eco-friendly bleaching agent as its degradation products are harmless, it is applied at alkaline pH and temperatures close to boiling are required for effective bleaching action which can cause fibre damage. A more specific process targeting only the coloured substances would be advantageous. If an enzyme-based system for bleaching can be developed, it can be integrated in the pectinase-based bioscouring process which has recently been implemented at an industrial scale. Laccase/mediator systems have been successfully used for the bleaching and modification of wood pulp fibres. Based on the observation that fungal laccases oxidize phenolic moieties of lignin in pulp and their capability of attacking phenolic hydroxyl groups, attempts have been made to use them to decolorize or eliminate flavonoids responsible for colour in cotton, thus resulting in bleaching action [41].

Laccases for bleaching denims to produce stone wash effect without the problem of back staining have become commercially available [42]. The laccase enzyme has also been used in combination with cellulase or after cellulase treatment to impart back-staining free washed look to denims [43, 44], as the use of cellulase alone causes back-staining due to redeposition of indigo dye [45].

Huge amounts of water and treatment with chemicals like thiosulphate are needed to remove hydrogen peroxide from fabrics which would otherwise cause problems in dyeing, especially in reactive dyeing where it can accelerate the dye hydrolysis. Catalase enzymes have been effectively used to clean up or remove the residual hydrogen peroxide in place of chemicals. Less water is required as rinsing steps are reduced, and the catalase-treated fabrics show uniform dyeing and good dye uptake [46]. Catalase-treated bleach cleanup bath itself can be subsequently used for dyeing [47].

Natural Dyes

Colouring matters derived from plant, animal and mineral resources were the only dyes available to man since prehistoric times till the first half of the nineteenth century. After the discovery of first synthetic dye in 1856 followed by many more, the traditional dyes being derived from natural sources were termed natural dyes. Subsequent rapid strides made in the field of synthetic dyes, their ease of application, ready availability in a wide range of colours, good colour fastness properties and suitability for use in large industrial set-up led to almost complete replacement of natural dyes with synthetic dyes for mainstream textile production.

However, the environmental awareness about the pollution caused by the use of synthetic dyes resulted in a global revival of interest in natural dyes since the last decades of twentieth century. The ban on certain azo dyes by Germany and other EU countries in 1995 in view of their being derived from suspected carcinogenic amines further boosted this interest which led to a rediscovery of natural dyes. Natural dyes are considered to be eco-friendly as these are renewable and biodegradable; are skin-friendly; and may provide health benefits to the wearer as many natural dye-yielding plant parts have been used as medicines in various traditional medicinal systems.

As the bulk of traditional knowledge in this area was lost due to many years of neglect, attempts were made to reconstruct the dye extraction and application processes by interacting with a few surviving traditional dyers and scanning whatever old records of this practice were available and by conducting research to know about the potential natural dye sources for textiles. It led to the publication of a number of books and articles about various natural dye sources and their application processes for textile fibres [48–58].

Natural dyes are usually not a single entity but a mixture of closely related chemical compounds whose content will vary depending upon the maturity, variety, and agroclimatic variations such as soil type and region. This makes the task of shade reproducibility difficult. Therefore, it is not possible to produce the same shade with a particular natural dye in every dyeing operation [59]. Even the differences in mineral content and pH of the water at two different places will produce a different shade with the same dye. That is a constraint in promoting large-scale use of natural dyes. The use of metallic mordants such as copper, tin and chrome for fixing and/or improving the colour fastness properties of natural dyes is not eco-friendly, and therefore, it is to be ensured that the content of restricted heavy metals in the dyed material is in compliance with the eco-regulations.

Dye-bearing natural materials contain only about 0.5–5 % dye; hence, these materials cannot be directly used in dyeing machinery as the large amount of plant matrix made up of a variety of non-dye constituents would result in patchy and uneven dyeing. This requires an additional dye extraction step to separate dyeing matter from non-dye plant matrix. Purified dye extracts commercially available now in countries such as USA, Europe, China and India are costly and are mainly used by hobby groups for the uniqueness of the shades. It is estimated that only about 1 % of the total textiles produced in the world are dyed by using natural dyes [60]. Traditional dyers, enthusiasts and hobby groups are the main users of natural dyes who work at cottage level.

Many natural dyes, if applied properly, can match the colour fastness properties of good synthetic dyes, but their colour range is rather limited with colours being soft and earthy. The most important advantage of natural dyes is their biodegradability. Bechtold et al. [49] have reported a reduction in pollution load with plant-based dyes in comparison with the use of synthetic dyes even with the latest dyeing techniques. Nayak [61] have found that the cost of dyeing textiles in blue, black and yellow shades with natural dyes is competitive with synthetic dyes when environmental cost of dyeing is also taken into account. Thus, the usage of natural dyes has the potential to reduce the risk of polluting the local water resources and offers a clean production model. The current dyestuff requirement from the industry being about 0.7 million tonnes [62] the use of natural dyes in mainstream textile processing is a big challenge as agricultural land is primarily required to feed an ever-increasing world population and support livestock. Also, biodiversity should not be compromised for the extraction of dyes, and therefore, in spite of being the natural choice for dyeing of organic textiles, the standard for such textile, GOTS, has banned the use of dyes and materials from endangered plants and has permitted the use of synthetic dyes having low environmental impact. Further, reasonable colour fastness levels are also required, and so only some natural dyes fulfilling these criteria can be used.

The use of agro and agroprocessing residues, microbial sources and cultivation of suitable dye plants in wastelands has the potential to enhance the availability of natural dyes for cottage and small-scale processors who can use these dyes for cleaner production to make their operations more sustainable as they do not have access to costly effluent treatment processes.

Use of Bioagents for Finishing

Use of Enzymes: Cellulase enzymes now find wide use in biopolishing of cotton textiles. Loose fibres adhering to the fabric which were previously removed by singeing can be removed by the treatment with cellulase. Thus, pilling can be reduced with a softer and smoother feel without the danger of fabric damage or yellowing due to exposure to the flame, and the fossil fuel used in singeing is saved. PET-based textile materials have been treated with esterase, lipase and cutinase enzymes to improve the moisture absorption characteristics as well as surface softness and to reduce pilling. Conventionally, the PET fabric has to be treated with sodium hydroxide to improve the above properties which led to higher strength loss and damage to the fabric. Enzymes have the potential to substitute the harsh chemical treatment and make the process eco-friendly [63].

Papain, a plant-origin proteolytic enzyme, has been used for shrink proofing of wool. However, the wool has to be first pretreated with a reducing agent such as sodium sulphite to achieve an effective shrink-resistant property.

Functional Finishing: Textile surfaces may act as nutrient for microbial growth which may cause unpleasant smells, staining, loss of mechanical strength and health-related problems for the wearer. An antimicrobial finish reduces the growth of microorganisms by either killing or inhibiting their growth through contact with the fabric surface [64]. Antimicrobial agents have been used on textiles since thousands of years ago; the use of spices and herbs as preservatives in mummy wraps by ancient Egyptians is an example [65]. Natural dyes and mordants such as myrobalan and turmeric used in earlier days for textile dyeing also possessed antibacterial properties in addition to providing colour. Antibacterial activity of many natural dyes can be attributed to the presence of tannins which have been reported to have antimicrobial properties against several strains of bacteria through in vitro studies [66–68]. Textiles dyed with such materials are also likely to show antimicrobial properties, and the same has been reported by many researchers [69–71].

The antimicrobial efficacy of these natural substances therefore would depend upon the type of tannins and their concentration in the substrate. Antimicrobial action of tannins can be due to their binding action with the proteins and enzymes present in the cell wall of microorganisms, thus inhibiting their metabolism and growth. Their capacity to bind with vital metal ions required by the microorganisms may also be a factor for growth inhibition [72]. Tannins generally have higher efficacy against Gram-positive bacteria as compared to Gram-negative bacteria. Their efficacy as antimicrobial agents for textiles may get limited due to the problems in getting a durable build-up of the required inhibitory concentration on textile material. Neem extract, due to the presence of azadirachtin, a tetranor-triterpenoid, has also been found to impart good antibacterial properties to textiles; however, its durability was poor.

Durability of plant extract-based antibacterial finishes in general is poor due to their lack of affinity to textile substrates. Application of these extracts after microencapsulation to trap the active antimicrobial agent using modified starch, gum acacia, sodium alginate, etc., as wall materials resulted in improved wash durability [73, 74].

Chitosan, obtained by alkaline deacetylation of chitin, a waste product of crab and shrimp processing industry, has also been reported to be effective against many common bacteria though the efficacy may vary according to molecular weight and degree of deacetylation. Chitosan and its derivatives have been used by for imparting antibacterial properties to textiles [75–77].

UV-Protective Finishing: Many natural dyes absorb in the ultraviolet region, and therefore, fabrics dyed with such dyes should offer good protection from ultraviolet light. Improvement in UV protection characteristics of natural cellulosic fibres after treatment with natural dyes has been reported by various researchers [78–80]. It was observed by Grifoni et al. [81] that treatment with tannins during mordanting itself improved the UV protection of fabrics. Saxena et al. [82] also reported that tannin-rich pomegranate rind extracts showed strong absorption in the UV region, and cotton fabrics treated with these extracts were imparted excellent UV protection which was durable to washing.

Aroma and Other Finishing: Microencapsulated essential oils have been used for aroma finishing of textiles. Such textiles not only create a feeling of freshness to the users but also have medicinal properties depending upon the essential oil used.

Microencapsulated citronella oil has been used for providing mosquito-repellent properties to textiles [83]. Chrysanthemum oil nanoemulsion has been used for wash-durable mosquito-repellent treatment of nylon nets [84].

Basak et al. [85] have used banana pseudostem sap to impart flame-retardant property to bleached and mercerized cotton textiles which were earlier pretreated with tannic acid and alum.

4.2 Process Improvement and Optimization

Combining two or more processes in a single step makes the process more sustainable due to lower consumption of water, chemicals and energy. For example, desizing and scouring processes can be combined. Also, scouring and peroxide bleaching can be combined as both require alkaline conditions. It may be possible to combine desizing, scouring and bleaching steps by using peracetic acid for bleaching. Various approaches to improve wet processing operations to reduce their environmental footprints have been tried, and some of these developments are reported here. Optimization of various process parameters also results in better sustainability by judicious use of resources.

Reactive dyes are mainly used for dyeing of cellulosics and their blends due to their good performance and cost-effectiveness. But the large quantity of salt (sodium chloride or sodium sulphate) used in dyeing is a cause of concern as it results in very high TDS of effluent which cannot be reduced by conventional effluent treatment methods, and costly processes like RO and ultrafiltration are needed. Cationization of cotton has been suggested for salt-free or low-salt dyeing with reactive dyes. Cationic agents such as 1-amino-2-hydroxy-3-trimethylammonium propane chloride [86] and 3-chloro-2-hydroxypropyl trimethyl ammonium chloride—CHTAC [87] have been suggested for this purpose. Applications of cationic polymers such as dimethylamino ethyl methacrcrylate [88], polyepichlorohydrin dimethylamine [89], polyamide–epichlorohydrin resin [90], poly(4-vinylpyridine) quaternary ammonium compound [91], dendrimers [92] and amino-terminated polymers [93] have also been suggested to make cotton cationic. Chitosan and its derivatives such as O-acrylamido-methyl-N-[(2-hydroxy-3-trimethylammonium) propyl] chitosan chloride (NMA-HTCC) have also been used for imparting cationic charge to cotton for low-salt dyeing of cotton [94, 95].

Replacement of the inorganic salts with biodegradable organic salts has also been attempted. Prabhu and Sundarajan [96] found that sodium citrate can be used as an alternative to inorganic electrolytes for dyeing of cotton fabrics with reactive and direct dyes with satisfactory results and significant reduction in TDS. An organic salt trisodium nitrilo triacetate was used as an alternative to sodium chloride for reactive dyeing of cotton by pad-steam method [97]. Dyeing was satisfactory with low pollution load. Organic electrolytes such as sodium edetate [98], sodium oxalate [99] and polycarboxylic acid sodium salts [100] have also been proposed as an alternative for inorganic electrolytes for dyeing.

Processing method used also has an effect on the effluent load. Exhaust processes usually require a higher material-to-liquor ratio; therefore, more water, energy and chemicals are required, and amount of effluent generated is also higher, whereas processes can operate at lower material-to-liquor ratio and therefore are more efficient. Cold pad-batch process for reactive dyeing has significant sustainability advantages. In this semicontinuous method, fabric to be dyed is first padded with liquor containing a mixture of reactive dyes and alkali (sodium silicate or sodium carbonate). It is then covered with polythene sheets to prevent evaporation of water and stored onto rolls. After a batching period of 6–12 h, the material is washed with water and hydrolysed dyes are removed by soaping. As salts, lubricants, levelling agents, fixatives and defoamers are not used in the process, effluent load is very less. Energy consumption is very less as dyeing and fixing takes place at ambient temperature.

Digital printing is an innovative development which has considerably improved the sustainability. Automatic rotary screen printing technique is mostly used for the textile printing. Digital printing is an inkjet-based application of colourants onto textile materials. The concept of this recent development in the printing field of the textile industry was initially introduced by Dr Sweet in early 1980s. Today, inkjet printers have become very popular for printing on textile substrates. It is a clean technology because of a high degree of utilization of printing inks and minimum water and energy consumption during post-treatment. It has the following advantages:

- Unlimited colour sampling and very good fastness,
- Easy switch over of designs without stopping the machine, whereas in conventional printing, each design requires making of screens, adjustment of pattern and sample printing before new design can be taken up and therefore a long downtime,
- Good reproducibility of designs,
- No limitation of size of the repeats of designs.

However, there are some limitations of the digital printing process also such as slower printing speed, requirement of high-quality special inks and specialized pretreatment of textiles to get good print quality which have slowed down its adoption rate.

4.3 Developments in Textile Chemicals, Dyes and Auxiliaries

There has been a general enhancement of awareness about environmental impact of textile processing among all stakeholders. Textile chemicals, dyes and auxiliary manufacturers have also made consistent efforts to develop chemistries which are more eco-friendly and sustainable. Dyes which have better exhaustion and can be dyed at lower temperature can reduce dye and energy consumption. Initial reactive dyes for dyeing of cellulosics had only a single functional group; hence, a good amount of dye was getting hydrolysed by reacting with water resulting in dye wastage and highly coloured effluent. To overcome this problem, bi- and multifunctional reactive dyes were introduced resulting in better dye utilization and less dye in effluent. It is claimed that Avitera® SE 3 range of polyreactive dyes introduced by Huntsman for cellulosic fibres and their blends can reduce water and energy usage by 50 % and also the salt consumption by 20 % during the dyeing process.

Prereduced vat and sulphur dyes have been made available by dye manufacturers. This significantly reduces the pollution caused by the use of reducing agent during dyeing and also simplifies the dyeing process. Archroma has recently launched dyes based on natural waste materials such as almond husks.

Coming to auxiliaries too, fatty alcohol-based ethoxylates have been commercially introduced as biodegradable non-ionic detergents in textile processing to replace poorly degradable nonyl and octyl phenol ethoxylates used earlier.

Sugar-based reducing agents have been introduced to replace sodium hydrosulphite for dyeing with vat and sulphur dyes.

Water, oil and stain repellency to textiles was earlier being provided by fluorocarbon finishes based on C8 chemistries. Due to the environmental issues associated with these finishes, shorter-chain C6 chemistry products have been introduced. Recently, Huntsman has introduced Phobotex® 3 an advanced range of fluorine-free hydropolymers for providing rain protection and stain management properties suitable for a wide range of textile end uses: outdoor rainwear, active wear, career wear and technical fabrics such as tarpaulins, boat covers, outdoor furnishings and shower curtains.

4.4 Development in Textile Machineries

Many developments have taken place in machinery design which result in substantial reduction in consumption of water, energy and chemicals during textile wet processing and make it more sustainable. Today, in batch processing of textiles, there is an emphasis on low liquor ratio processing as a reduction in the amount of process water use per unit textile weight would not only save water, but also reduce the chemical usage, energy requirements and the quantity of generated effluent. Many types of machinery which use low and ultralow liquor ratios for processing of textile materials are now available from various machinery manufacturers. Processing machinery equipped with microprocessor-based controllers reduces energy consumption and CO_2 emission. Similarly, installation of an automated chemical dispensing system optimizes chemical use. Savings due to lower consumption of energy and chemicals ensure a short payback time.

Continuous bleaching and dyeing ranges (CBRs and CDRs) have been introduced by machinery manufacturers and adopted by the processing industry as continuous processing is more efficient than batch processing. The latest CBRs are equipped with prewashers, dosing system with automatic controls, combi steamer, efficient washing units and dryers. Continuous, efficient, counter-current flow washers reduce effluent volume. Open width washers based on continuous interchange of water around fabric with lower contamination wash waters with built-in spray, multinips, vacuum extraction and ultrasonic technology drastically reduce water and steam consumption and provide a highly concentrated effluent for recycling. These also have smaller space requirements and lower operational costs [101].

Vacuum application of dyes; high-efficiency padders; and other dye application systems for continuous dyeing have reduced the environmental impact of the dyeing process. Vacuum dyeing systems for smaller lots have also been developed where dyebath size can be reduced to less than 15 L from 140 to 150 L required for conventional exhaust dyeing systems such as winch. Air flow dyeing technology is an improvement over soft flow dyeing system where liquor ratios of 1:3 can be used and overflow liquor mixed with air in mist form moves the fabric. It is therefore a very gentle process highly suitable for processing of delicate fabrics such as knits. Closed HTHP jiggers have been developed which can be used for dyeing of PET in open width using ultralow liquor ratios of 1:2.

e-control dyeing process introduced by Monforts in collaboration with DyStar at ITMA in 1995 slightly modifies the reactive dyeing process as both, drying and fixation, take place at the same time. The e-control climate inside the fixation chamber ensures a perfect dyeing result during the drying process. By using this process, cotton, viscose, tencel, and linen can be dyed. It is claimed that this process uses less energy, water and chemicals than conventional processes [102].

High-speed stenters with self-lubricating chains requiring very low maintenance have been developed for textile finishing operations. Process control and automation for high energy efficiency have been introduced. Cleaning of exhaust air is undertaken to reduce pollution, thereby making stenters eco-friendly.

Insulation of steam pipes and machinery conserves energy and makes the environment less hot for the workers. The use of heat exchangers can conserve energy usage in wet processing operations and can also bring down the temperature of the effluent.

5   Developments in Recycling and Reuse of Process Inputs

Recycling and reuse of process inputs can reduce the costs by economizing on the quantity of inputs to be procured and also reduces the environmental impact. Advanced membrane processes such as microfiltration, ultrafiltration, nanofiltration, RO, advanced oxidation processes, electrochemical processes, adsorption and ion exchange processes are quite effective for the removal of colour and COD from textile wastewater and appear promising in terms of their performance and cost for treatment and reuse of textile effluents.

5.1   Recovery and Reuse of Process Chemicals

As textile processing baths usually contain a number of chemicals and auxiliaries in dilute solutions or dispersions, their individual recovery is difficult and may not be cost-effective. Also they may undergo changes during the process. Therefore, only a few instances of recovery and reuse of process chemicals are found in textile wet processing. It is common knowledge that alkali containing mercerizing wash liquor can be used for scouring and bleaching. Also, recovery of synthetic sizes like PVA from process waters through membrane filtration techniques is being used by the industry to improve the process economy and reduce the pollution load. As dyes remaining unused in supercritical fluid dyeing process are recovered in pure form, these may be reused for dyeing fresh samples.

5.2   Water Recovery and Reuse

As textile chemical processing operations are carried out in aqueous medium, water is the input which is used in larger quantities. Clean soft water is a precious commodity. Therefore, attempts have been made to reuse the treated effluent for various wet processing operations as that reduces the demand for freshwater. Recovery and reuse of water is beneficial both for conserving and supplementing available water resources and for reducing the environmental pollution. Water recycling and reuse is a necessity for implementing zero liquid discharge system now being promoted due to environmental concerns. Efficiency and cost economics for recycling and reuse of water would, however, depend upon the process parameters, chemicals and machinery used.

Lu et al. [103] in a study conducted in a 600 m^3/day pilot plant observed that the average removal efficiencies of COD, colour and turbidity from wastewater were about 93, 94.5 and 92.9 %, respectively, by using biological treatment and membrane technology. Treated water had COD value below 50 mg/L, no suspended solids and acceptable values for colour and turbidity, and its quality was satisfactory for dyeing and finishing process except for dyeing light shades. Operating cost for wastewater reclamation was approximately 0.25 US$/m^3, and thus, wastewater reclamation and reuse was found to be quite promising. Lanza [104] has reported that in a carpet dyeing and finishing plant using approximately 150,000 gallons of water per day, installation of chemical treatment and polymer filter disks for treatment of process effluent could result in $300 per day savings by reusing treated water for processing and the savings were sufficient to recover the initial installation costs within a year.

6   Use of Waterless Technologies for Pollution-Free Textile Processing

Wet processing as the name suggests is carried out in aqueous medium. Some new revolutionary developments intend to carry out these operations without using water as a medium so that the pollution issues can be avoided. The use of supercritical fluids as processing medium and use of plasma are two such emerging technologies discussed in the following sections though the use of laser technology to achieve a faded look and worn-out effect on denim materials is also a waterless technology. Special novel effects not possible earlier have now been produced by combining computer designing and laser engraving. Varied degree of colour removal with little or no damage to the other properties of denim material can be achieved by using different laser parameters. As it is a waterless process and has high potential for innovation, it has an edge over other conventional processing techniques [105–107].

6.1 Supercritical Fluid Technology

A gas is a supercritical fluid above its critical values of temperature and pressure where distinct gas and liquid phases do not exist. Such fluids have physical properties somewhere between those of a liquid and a gas. They are able to spread out along a surface more easily than a true liquid because they have much lower surface tension than liquids. As their viscosity is also low, they have very good diffusivity and thus better interaction with the substrate. Critical temperature and pressure values for carbon dioxide (CO_2) are 31.4 °C and 1070 lb per square inch (psi) or 73.8 bars, respectively, which are much lower than those for many substances. It is non-toxic, non-inflammable, cheap and easily available and does not leave residues. Being a non-polar molecule, it behaves like a non-polar organic solvent in its supercritical state. Supercritical carbon dioxide (SC CO_2) has been employed for extraction of high-value compounds from natural substances, especially for food applications as unlike organic solvents; there is no residual solvent in the extracted material.

SC CO_2, due to its non-polar nature, can easily dissolve water-insoluble dyes such as disperse dyes without using any dispersing agent. Materials dyeable with dyes, such as PET, polypropylene (PP) and poly lactic acid, which are problematic to dye in aqueous medium, can be dyed easily using SC CO_2. Though the concept of SC CO_2 dyeing was proposed in 1980s, its use for practical dyeing applications started only in the last decade of the twentieth century.

These dyeing systems operate at high temperature and pressure and basically consist of an HTHP vessel, CO_2 tank, dye container, and a compressing and a circulation pump. The material to be dyed is placed in the HTHP vessel. Then, the system is brought to the required pressure and temperature to bring CO_2 into supercritical state which then dissolves the dye. The resultant dye liquor is circulated in and out of the HTHP vessel by the circulating pump. After completion of dyeing, dye solution is depressurized which brings the CO_2 into gaseous state leaving behind the dyes. Both CO_2 and dyes are collected and reused. Pure SC CO_2 is then circulated in the HTHP vessel to remove the unfixed dyes from the dyed material. Thus, no water is used and there is no effluent generation.

Most work on dyeing using SC CO_2 has been carried out on PET and has come up to the commercial level. The operating conditions for the PET dyeing range from 60 to 150 °C and 100 to 350 bar pressure [108]. Heat setting prior to dyeing is recommended to avoid strength reduction and shrinkage. It is also useful for dyeing of PP fibres as their high crystallinity and non-polar nature create problems in aqueous dyeing. These can be dyed with good all-round fastness properties by using SC CO_2 system. Polylactic acid (PLA) fibre in spite of its sustainable nature has found limited application in textile industry due to loss in mechanical properties during traditional aqueous-based processing. These fibres could be successfully dyed using disperse dyes in SC CO_2 medium with good retention of mechanical properties [109].

The SC CO_2 dyeing process is highly advantageous over the conventional aqueous-based systems from the sustainability viewpoint. Both dyes and CO_2 used as solvent in its supercritical state are recovered; there is no effluent generation and therefore no adverse effect on the environment. The cost incurred for treating the effluent is saved as also the energy required for drying of dyed material as the solvent used is a gas under normal atmospheric conditions. There is a complete saving of about 100–150 L water which is normally required to dye one kg of PET.

Dutch company, DyeCoo Textile systems, was the first to launch commercial SC CO_2 dyeing system. It has partnered with Huntsman Textile Effects to jointly develop supercritical CO_2 textile processing technology. Adidas was the first brand to introduce a product line using this technology with the manufacturing of 50,000 drydye T-shirts in 2012 [110]. This technology has also been found useful in wax removal/scouring of cotton textiles [111].

In spite of the many sustainability advantages, high equipment cost has restricted the widespread adoption of supercritical CO_2 technology by the textile industry. Poor dyeability of natural fibre and man-made cellulosic textiles in SC CO_2 due to poor solubility of polar dyes in SC CO_2 and poor affinity of non-polar disperse dyes towards these fibres have also restricted its use. Different approaches such as fibre modification, dye modification and adding of a cosolvent or a modifier to improve the solubility of slightly polar solutes have been attempted to solve this problem but have not achieved much success, and therefore, more innovations are required for better adoption of this technology.

6.2 Plasma Technology

Plasma is generated when a substance in its gaseous state absorbs high energy and gets ionized into positively charged atoms and free electrons. Also termed the fourth state of the matter, it consists of positive and negative ions, electrons, neutrals, radicals and photons. It is thus highly reactive and can easily react with many substances. Existing chemical bonds on the surface are broken, and new bonds are formed in the process, thereby introducing new functionalities. Besides, some physical changes on the surface may also take place. These physicochemical changes take place at surface only, and the bulk remains largely unaffected. It has therefore many applications in textiles where surface properties are responsible for various end use properties of textile products such as wettability, dyeability, printability, felting shrinkage in case of wool, pilling, electrostatic properties and water resistance. It may be possible to obtain surface characteristics to meet specific requirements by appropriate selection of plasma composition, i.e. selection of gases one or more from O_2, N_2, H_2, air, Ar, He, NH_3, hydrocarbons and fluorocarbons, and process conditions such as treatment time, power, pressure and gas flow rate.

Plasma can be of two types: hot or thermal plasma where temperature of the plasma zone exceeds 1500 °C, and cold or non-thermal plasma where temperature of the plasma zone is close to ambient temperature. Thermal plasma finds application in metal, electrical and materials industry but is not suitable for use in textiles as textile materials which are polymeric in nature get degraded at higher temperatures. Therefore, it is the cold plasma which is used for carrying out modifications in textiles. Energy to a gaseous medium for generation of plasma can be supplied through several means such as chemical, radiant, nuclear, thermal ionization at high temperatures, or electrical discharges by applying an electrical potential, but electrical breakdown of gaseous molecule in the presence of sufficient discharge voltage across the electrodes with high-frequency AC signal is the most popular and commonly used method for generating plasma.

Plasma can be further categorized into low-pressure or vacuum plasma and atmospheric pressure plasma (APP) depending upon the pressure inside the plasma reactor. Low-pressure plasma was developed many years ago as it is easy to ionize a gaseous molecule by electrical breakdown under low-pressure conditions. It has been extensively studied for material processing as also for textile applications. However, it has not found commercial applications in textiles and allied industries due to economic considerations as to develop and maintain low-pressure conditions in the big-size commercial reactors is costly, and moreover, being a batch process, it is not suitable for integration into the current continuous textile processing machinery. Though it is difficult to generate and stabilize the plasma at atmospheric conditions, but if it can be achieved, it can offer a cost-effective and faster processing technology, which can be easily integrated with the existing textile processes. Hence, atmospheric pressure cold plasma has received much interest among researchers as well as industry as an alternative sustainable processing technique for varied applications in textiles.

As the reactants are in gaseous form, plasma processing is a dry process, and hence, there is no effluent generation. It is also a highly energy-efficient and clean process. This leads to saving a large quantity of water, chemicals and electrical energy while avoiding the production of large volumes of effluent or hazardous by-products. The various sustainability benefits of the plasma treatment can be summarized as follows:

- Dry process, no water required for processing other than for cooling of reactor,
- Lower requirement of chemicals in comparison with conventional processing,
- No need to dry processed textile,
- Lower energy consumption,
- Faster processing speed,
- Versatile process, possibility to process different types of products using the same reactor by changing the gases used for plasma generation,
- Only surface changes and no damage to bulk.

Value-added functionalities such as water, stain and oil repellent, hydrophilic, removal of size, natural waxes and also improvements in dyeing, printing, biocompatibility and adhesion can be achieved by modifying the fibre surface at nanometre level through plasma. Surface modification with a desired functionality can be produced by selecting the appropriate plasma parameters and ratio of carrier to precursor molecules. Fragmentation of precursor molecules followed by the reaction of plasma with the textile fibres is an effective tool for nanoscale surface engineering of textiles to develop advanced, technical, apparel and home textiles. Plasma reaction at atmospheric pressure is challenging due to the presence of high-density ions, electrons, excited particles and ready availability of reaction inhibitors such as atmospheric air and oxygen.

Various uses of plasma in textiles are discussed below:

Textile preparatory processing

Plasma can be used to assist in removal of the contaminants, finishing and sizing agents from the textile materials. PVA is primarily used for slashing synthetic yarns and may be used as a secondary sizing agent to starch for cotton yarns. A complete PVA size removal is difficult and requires high energy and water consumption as it is soluble in hot water. Atmospheric plasma treatment has been found to increase the cold water solubility of PVA on cotton [112]. Plasma treatment has been reported to be used by Kan et al. [113] for removal of starch sizes from cotton and for reducing the time for cotton scouring [114]. It has been reported that plasma treatment can significantly reduce soap concentration and washing temperature in degumming process of silk [115].

Plasma treatment has been used for making the synthetic fibres such as PET and PP hydrophilic [116]. Oxygen plasma is usually suitable for such applications. It can reduce the wetting time of these substrates and increase their antistatic and adhesion property. Surface etching produced by plasma treatment greatly improves the inkjet printability of PP.

Improvement in coloration

Plasma treatment has been reported to increase the exhaustion of dyes on textiles, especially on protein fibres such as wool [117]. In the studies conducted by Teli et al. [118] on silk also, almost complete exhaustion of acid dye was achieved at a much lower temperature and in shorter time duration. This can be of much use in improving the dyeing process sustainability. However, results on cellulosic fibres are not very conclusive, and in most cases, plasma treatment has not resulted in any appreciable change in colour strength. Man et al. [119] have, however, reported an improvement in colour yield, levelness and rubbing fastness of red pigment dye on cotton with oxygen plasma treatment.

APP pretreatment has been reported to significantly increase the colour yield of the digital inkjet-printed cotton fabrics [120]. Colour fastness to washing and rubbing and outline sharpness were also improved in comparison with the control cotton fabric printed without APP pretreatment.

Applications in textile finishing

Many studies on plasma treatment to achieve antifelting properties on wool have been reported. Felting of wool occurs primarily because of the specific scalelike, hydrophobic surface of fibres. Plasma treatment induces morphological and frictional changes which reduce the felting shrinkage of wool. However, plasma treatment alone is not satisfactory in terms of durability and results in a rough handle, and therefore, subsequent treatment with a suitable polymer is necessary. The resultant product after application of polymers is machine washable, and the effect is achieved without using chlorine. Pretreatment of wool with air/oxygen plasma followed by coating with polyurethane-based resin has been found to impart good antifelting properties [121].

Another main application of plasma in textiles had been for producing water-, stain- and oil-repellent properties. These features are achieved by the use of plasma containing fluorine molecules. If a fluoroalkane such as tetrafluoromethane or hexafluoroethane is used as a process gas, fluorine gets substituted for abstracted hydrogen on the surface of the substrate, reducing its surface energy and making it oil and water repellent. This process has been carried out on varied textile substrates such as PET, nylon and cellulosic materials [122–124]. The advantage of the process is that there is no change in the comfort properties of textiles such as air permeability due to the nanoscale of the process and the process is completed in a single step. No further drying and curing is required; thus, there is a saving of energy, time and chemicals.

Plasma processing machinery
Low-pressure plasma, requiring closed, vacuum system equipment, was unsuitable for textile processing on an industrial scale as capital equipment cost as also the operational and maintenance costs would be high. APP offers an alternative, attractive, low-cost, environmentally friendly route for textile manufacturing, with improved quality and performance. Though plasma processing of textiles is still in an experimental stage and a number of issues need to be addressed before it becomes a commercial alternative, a number of manufacturers have developed pilot- to commercial-scale machinery for plasma treatment of textiles for various applications, and some of these manufacturers are listed in Table 5.

Table 5 Plasma machinery manufacturers

Manufacturer	Application
Acxys, France	Wettability, water repellent (www.acxys.com)
Apjet, USA	Water and stain repellency (www.apjet.com)
Arioli, Italy	Water repellent (www.arioli.biz)
Diener, Germany	Cleaning, etching, activation, polymerization [125]
Dow corning corporation	Surface modification and coating [126]
Europlasma, Belgium	Surface coating, water repellent (www.europlasma.be)
Grinp, Italy	Surface modification and coating (www.grinp.com)
Plasmatreat, USA	Self-cleaning, flame retardancy (www.plasmatreat.com)
Sigma, USA	Surface modification and coating [127]
Softal, Germany	Water repellent, wettability (www.softol.de)
Vito, Belgium	Cleaning, activating, coating (www.vitoplasma.com)

7 Role of Eco-standards and Environmental Regulations in Promoting Sustainability

Eco-labels by governmental or independent bodies aim to set environmental standards for products. Concern about the environment among consumers especially in EU countries has been growing in recent years, and more and more people are interested in buying green alternatives to regular products. However, just by looking at a textile product, it is not possible to know whether it is eco-friendly. An eco-label identifies the general environmental performance of a product spanning its entire textile supply chain and contributes to consumer safety and reducing the environmental impact, thereby supporting sustainable textile consumption patterns. Government, industry, commercial associations, retailers, companies and consumers are all major participants in the scheme. Eco-labels were first introduced during the last decade of the twentieth century after the German ban on azo dyes derived from amines suspected to be carcinogenic, and today, a number of regional as well as global eco-labels are in existence. These labels cater to environment conscious customers by ensuring the quality and performance of products as well as usage of materials that are safe for human health and environment during their manufacturing. Manufacturers must meet certain requirements before their products can be classified to be 'green'. Participation in an eco-label scheme is voluntary; companies submit their products for third-party compliance testing and/or verification to obtain an eco-label for specific products that meet detailed established environmental guidance criteria. That eco-label may be then displayed on the products that meet these requirements. Widely recognized eco-labels are helpful guidelines for consumers who want to buy eco-friendly sustainable products.

These labels promote the communication of authentic and verifiable information on environmental aspects of products and services, encourage the demand and supply of products and services that cause less strain on the environment and boost market-driven efforts for sustainable manufacturing. When a product is approved by an eco-label, permission to use the scheme's distinctive eco-label or symbol is granted for a specified period. The award is periodically reviewed to ensure that they comply with the evolving criteria, technological developments and market advances. Some examples of the criteria specified by these labels are as follows: textile products should contain limited amounts of substances harmful to health and the environment and should be processed with reduced use of water and air pollution. Laundry detergents should not contain certain substances considered to be harmful to the environment or which promote the growth of algae in water bodies; be mostly biodegradable; and include ecological washing instructions. The Oeko-Tex 100 (product certification), Oeko-Tex 1000 (factory certification) and the Bluesign® International Standards are examples of eco-labels that attempt to provide clear information on the impact of textile products on people and the environment. The Bluesign® system requires the entire textile supply chain to jointly reduce its impact as its five principles, viz. resource productivity, consumer safety, water emission, air emissions and occupational health and safety cover the entire value chain.

REACH standing for Registration, Evaluation, Authorization and Restriction of Chemicals is a regulation of the European Union which came into force in 2007. By mandating the registry of details of chemicals manufactured or imported into EU in amounts exceeding the prescribed limit, it aims to improve the protection of human health and the environment from the risks posed by chemicals across various sectors. The use of registered chemicals is authorized or restricted according to their hazard and risk data.

The OEKO-TEX® Association has recently introduced the new ECO PASSPORT certification for sustainable textile chemicals. It is a two-step verification procedure for textile processing chemicals, colourants and auxiliaries which enables the manufacturers to confirm that their products meet the specific criteria for environmentally responsible textile production [128]. EU official label for green textiles— EU Flower limits the amount of toxic residues in textile materials, limits the level of metallic impurities in dyes and pigments, prohibits the use of azo dyes that cleave to a list of aromatic amines or dyes classified as carcinogenic, mutagenic and toxic for reproduction and also specifies the pH and temperature and COD content of the textile wastewaters.

The latest 4.0 version of the Global Organic Textile Standard (GOTS) introduced in March 2014 sets requirements from the harvesting of raw materials to manufacturing and labelling to assure the consumers of the organic status of textiles. However, it takes into consideration that industrial-scale production of textiles is not possible without the use of chemicals. Hence, at each manufacturing stage, a list of materials is provided which are safe and allowed and which are not allowed for use. Minimum impact on the environment, minimum hazard and toxicity are the criteria for allowing the use of a material. Therefore, natural dyes and other chemicals from endangered plant species are not permitted, but synthetic dyes which can meet toxicity and hazardous substance criteria are allowed. Minimum colour fastness requirements have also been specified. The objective is to ensure good-quality textile products not containing harmful substances which are produced in an environmentally and socially responsible manner, thus ensuring sustainability.

Almost all the eco-labels have set the limits for the harmful substances such as heavy metals, banned carcinogenic amines, brominated and chlorinated flame retardants, APEO and formaldehyde release in the finished textile products and put emphasis on the sustainable manufacturing process while ensuring the minimum social requirements.

8 Social and Economic Issues

Social and economic aspects are an integral part of sustainability of any process. Textile processing has to be economically competitive, but at the same time, social issues also need to be considered. As the long textile value chain involves a number of players from diverse sectors such as textile fibre growers, spinners, chemical suppliers, processors, garment makers, transporters, retailers and consumers, these issues are of concern for each of them and every sector is responsible in ensuring sustainability of textile manufacturing. Regarding the chemical processing sector, as discussed in the previous section, many voluntary eco-labels introduced for textiles look at a product in a holistic way and are concerned about its production process as well which should have low environmental impact. They look for consumer safety as also the occupational safety of workers during manufacturing and insist upon payment of fair wages and thus are contributing to the promotion of sustainable textile production. The low-impact sustainable textile manufacturing processes need wider adoption as the environmental impact caused by these processes are endangering the water resources in many developing and undeveloped countries and thereby the well-being and livelihoods of communities residing in surrounding areas dependent on in these resources. Many textile industries in these countries are not willing to adopt sustainable chemical processing options as they are interested in quick profits and also lack the required financial and technical resources to implement the low-impact technologies. It would require investment and initial costs would be higher for adopting sustainable technologies, but in the long run, savings in energy, chemicals and process water consumption would result in lower costs. These would appear still more attractive if the environmental damage and cost of restoring it are taken into account. Governments and law makers therefore have to contribute by putting stricter regulations and their proper implementation for the protection of environment. At the same time, they should also offer incentives and support to processors for adoption of low-impact technologies. Big brands and companies buying these textiles as also the consumers have a social responsibility to minimize the damage to environment and should be ready to pay extra if required to promote sustainable textile production. Environmental issues are of concern to society at large as greenhouse gas emissions, and the resultant climate change would affect everyone. Also the air and water pollutants can affect much larger population by spreading through wind, rain and consumption of contaminated food products. Therefore, all stakeholders in the sector need to come together and contribute to ensure sustainability. Further research efforts from chemical manufacturers, machinery manufacturers and researchers should continue to further develop low-cost low-environmental-impact products for minimizing the adverse social and economic implications of textile wet processing.

9   Future Outlook and Conclusion

As discussed in this chapter, quite a number of solutions are available for various textile wet processing operations which can significantly reduce the detrimental impact on the environment. However, there are still many challenges and many promising technologies such as plasma and supercritical CO_2 processing need further development and refinement to make them a viable alternative. Similarly, efforts are needed for developments in the field of biodegradable dyes, chemicals and auxiliaries as also enzyme assisted processing in view of the environmental advantages. It is the responsibility of the concerned industry and researchers to further develop viable, minimum environmental impact options.

Adoption of the available best practices for sustainable textile wet processing has been poor as most of the operations are located in developing and undeveloped countries in decentralized sector and these units lack the resources to install and follow state-of-the-art facilities and practices. Governments therefore have two roles to play in such scenario. First, they need to put stricter regulations and ensure their proper implementation in order to promote sustainable textile production and to ensure clean environment and livelihoods for the people. Secondly, along with investors, they should provide financial support to help in modernizing the sector and reducing its environmental footprints. Voluntary eco-labels and certification schemes have contributed towards achieving these objectives in a limited way.

Each player across the sector has to work towards ensuring sustainability. Certain non-profit and trade organizations such as Textile Exchange and Sustainable Apparel Coalition involving all stakeholders are now working to reduce the environmental and social impacts of textile production around the world. The sector as a whole has to aim towards achieving zero discharge of hazardous chemicals and draw a road map to reach this goal. This is urgently needed to prevent the further contamination of water bodies with hazardous substances and the resultant build-up in people and animals. Toxic pollution has to be dealt with in all countries as otherwise the pollution of water resources along with global warming caused due to the release of greenhouse gases and associated climate change would threaten our livelihoods and our future. Brands, retailers and consumers on their part need to be environment conscious and be ready and willing to pay more for the products which do not harm people and the environment.

References

1. Anon (2015) 30 shocking figures and facts about global textile and apparel industry. http://www.business2community.com/fashion-beauty/30-shocking-figures-facts-global-textile-apparel-industry-01222057#ADLVmKPEeCsdUlzQ.99. Accessed 8 Mar 2016
2. Greenpeace International (2011) Dirty laundry: the toxic secret behind global textile brands http://www.greenpeace.org/international/Global/international/publications/toxics/Water%202011/dirty-laundry-12pages.pdf. Accessed 28 Mar 2016
3. Greenpeace International (2015) Footprints in the snow. http://www.greenpeace.org/international/Global/international/publications/toxics/2015/Footprints-in-the-Snow-Executive-Summary-EN.pdf. Accessed 12 May 2016
4. Ramesh Babu B, Parande AK, Raghu S, Prem Kumar T (2007) Textile technology: cotton textile processing waste generation and effluent treatment. J Cotton Sci 11(3):141–153
5. Naveed S, Bhatti I, Ali K (2006) Membrane technology and its suitability for treatment of textile waste water, Pakistan. J Res (Sci) 17(3):155–164
6. Wasif AI, Indi IM (2010) Combined scouring and bleaching of cotton using potassium persulpahte. Indian J Fibre Text Res 35:353–357

7. Karamkar SR (1999) Textile science and technology, chemical technology in the pre-treatment processes of textiles. Elsevier, Amsterdam
8. Anon (2013) APEOs and NPEOs in textiles in O ecotextiles (on line) https://oecotextiles. wordpress.com/2013/01/24/apeos-and-npeos-in-textiles-2/. Accessed 12 Aug 2014
9. GG 62 (1997) Water and chemical use in the textile dyeing and finishing industry. Guide produced by the environmental technology best practice programme, UK. Available at http://www.wrap.org.uk/sites/files/wrap/GG062.pdf. Accessed 5 May 2016
10. Zaharia C, Suteu D, Muresan A, Muresan R, Popescu A (2009) Textile wastewater treatment by homogenous oxidation with hydrogen peroxide. Environ Eng Manage J 8(6):1359–1369
11. Shukla SR (2007) Pollution abatement and waste minimization in textile dyeing. Environ Aspects Text Dyeing, pp 116–148
12. Kissa E (2001) Fluorinated surfactants and repellents. Surfactant science series. Marcel Dekker, New York, p 97
13. Buck et al (2011) Perfluoroalkyl and Polyfluoroalkyl substances in the environment: terminology, classification, and origins. Integr Environ Assess Manage 7:513–541
14. Darnerud PO, Eriksen GS, Jóhannesson T, Larsen PB, Viluksela M (2001) Polybrominated diphenyl ethers: occurrence, dietary exposure, and toxicology. Environ Health Perspect 109 (Suppl 1):49–68
15. Shin JH, Baek YJ (2012) Analysis of polybrominated diphenyl ethers in textiles treated by brominated flame retardants. Text Res J 82(13):1307–1316. doi:10.1177/0040517512439943
16. Rajkumar AS, Nagan S (2011) Study on Tiruppur CETPs discharge and their impact on Noyyal river and Orathupalayam dam, Tamil Nadu (India). J Environ Res Dev 5(3):558–565
17. Uqaili MA, Harijan K (eds) (2011) Energy, environment and sustainable development. Springer Science & Business Media, New York
18. World Bank (2010) A detailed analysis on industrial pollution in Bangladesh. Workshop Discussion Paper Transfer, EPA-625/3- 74-004, World Bank Dhaka Office, Dhaka, Bangladesh
19. Mansour HB, Houas I, Montassar F, Ghedira K, Barillier D, Mosrati R, Chekir-Ghedira L (2012) Alteration of in vitro and acute in vivo toxicity of textile dyeing wastewater after chemical and biological remediation. Environ Sci Pollut Res 19(7):2634–2643
20. Solmaz SKA, Birgul A, Ustün GE, Yonar T (2006) Colour and COD removal from textile effluent by coagulation and advanced oxidation processes. Color Technol 122(2):102–109
21. Gomez N, Sierra MV, Cortelezzi A, Rodrigues Capitulo A (2008) Effects of discharges from the textile industry on the biotic integrity of benthic assemblages. Ecotoxicol Environ Saf 69:472–479
22. Lloyd B, Helmer R (1992) Surveillance of drinking water quality in rural areas, longman scientific and technical. Wiley, New York, pp 34–56
23. Kumar V, Chopra AK, Chauhan RK (2012) Effects of textile effluents disposal on water quality of sub canal of Upper Ganga Canal at Haridwar (Uttarakhand), India. J Chem Pharm Res 4(9):4206–4211
24. Odjegba VJ, Bamgbose NM (2012) Toxicity assessment of treated effluents from a textile industry in Lagos, Nigeria. Afr J Environ Sci Technol 6(11):438–445
25. Hasan MK (2014) Impacts of textile dyeing industries effluents on surface water quality: a study on Araihazar Thana in Narayanganj District of Bangladesh Md. Khalid Hasan, Mahin Miah. J Environ Human 1(3). ISSN(Print): 2373–8324 ISSN(Online): 2373–8332 doi:10. 15764/EH.2014.03002
26. Mehari AK, Gebremedhin S, Ayele B (2015) Effects of Bahir Dar textile factory effluents on the water quality of the Head Waters of Blue Nile River, Ethiopia. Int J Anal Chem 2015:7. Article ID 905247. http://dx.doi.org/10.1155/2015/905247. Available on http://www. hindawi.com/journals/ijac/2015/905247
27. Muthu SS, Li Y, Hu JY, Ze L (2012) Carbon footprint reduction in the textile process chain: recycling of textile materials. Fibers Polym 13(8):1065–1070

28. Sherburne A (2009) Achieving sustainable textiles: a designer's perspective. In: Blackburn RS (ed) Sustainable textiles: life cycle and environmental impact. Woodhead Publishing, Oxford pp 3–31
29. Sharma S (2012) Energy management in textile industry. Int JPower Syst Oper Energy Manage ISSN (PRINT) 2(1):2
30. Kocabas AM (2008) Improvements in energy and water consumption performances of a textile mill after BAT applications (Doctoral dissertation, Middle East Technical University). Available from https://etd.lib.metu.edu.tr/upload/12609296/index.pdf. Accessed 12 May 2016
31. Hasanbeigi A (2010) Energy-efficiency improvement opportunities for the textile industry. http://www.energystar.gov/sites/default/files/buildings/tools/EE_Guidebook_for_Textile_industry.pdf. Accessed 12 Aug 2014
32. Ströhle J, Schramek G (2016) Textiles achieve ecological footprint, new opportunities for China. http://www.benningergroup.com/uploads/media/Carbon_Footprint_China_EN.pdf. Accessed 28 Mar 2016
33. Zaffalon V (2010) Climate change, carbon mitigation and textiles. Textile World. Available online http://textileworld.com/Articles/2010/July/July_August_issue/Features/Climate_Change_Carbon_Mitigation_In_Textiles.html. Accessed 12 Aug 2014
34. Kant R (2012) Textile dyeing industry an environmental hazard. Nat Sci 4(1):22
35. Arputharaj A, Raja ASM, Saxena S (2015) Developments in sustainable chemical processing of textiles. Green fashion. Springer, Singapore, pp 217–252
36. Saravanan D, Sivasaravanan S, Sudharshan Prabhu M, Vasanthi NS, Senthil Raja K, Das A, Ramachandran T (2012) One-step process for desizing and bleaching of cotton fabrics using the combination of amylase and glucose oxidase enzymes. J Appl Polym Sci 123:2445–2450
37. Eren HA, Anis P, Davulcu A (2009) Enzymatic one-bath desizing—bleaching—dyeing process for cotton fabrics. Text Res J 79(12):1091–1098
38. Kim J, Kim SY, Choe EK (2006) The beneficial influence of enzymatic scouring on cotton properties. J Nat Fibers 2(4):39–52
39. Menezes E, Choudhari M (2011) Pre-treatment of textiles prior to dyeing. INTECH Open Access Publisher
40. Hebeish A, Ramadan M, Hashem M, Shaker N, Abdel-Hady M (2009) New development for combined bioscouring and bleaching of cotton-based fabrics. Res J Text Apparel 17(1):94–103
41. Pereira L, Bastos C, Tzanov T, Cavaco-Paulo A, Gübitz GM (2005) Environmentally friendly bleaching of cotton using laccases. Environ Chem Lett 3(2):66–69
42. Rodríguez-Couto S (2012). Laccases for denim bleaching: an eco-friendly alternative. Sigma 1:10-2
43. Montazer M, Maryan AS (2010) Influences of different enzymatic treatment on denim garment. Appl Biochem Biotechnol 160(7):2114–2128
44. Tarhan M, Sarıışık M (2009) A comparison among performance characteristics of various denim fading processes. Text Res J 79(4):301–309
45. Pazarlioğlu NK, Sariişik M, Telefoncu A (2005) Treating denim fabrics with immobilized commercial cellulases. Process Biochem 40(2):767–771
46. Amorim AM, Gasques MD, Andreaus J, Scharf M (2002) The application of catalase for the elimination of hydrogen peroxide residues after bleaching of cotton fabrics. Anais da Academia Brasileira de Ciências 74(3):433–436
47. Tzanov T, Costa S, Guebitz GM (2001) Dyeing in catalase treated bleaching baths. Color Technol 117(1):1–5
48. Adrosko RJ (1971) Natural dyes and home dyeing. Dover, New York 160p
49. Bechtold T, Turcanu A, Ganglberger E, Geissler S (2003) Natural dyes in modern textile dyehouses—how to combine experiences of two centuries to meet the demands of the future? J Clean Prod 11(5):499–509
50. Bhattacharyya N (2010) Natural dyes and their eco-friendly application. IAFL, New Delhi 308 p

51. Buchanan R (1987) A weaver's garden: growing plants for natural dyes and fibers. Dover, New York 228p
52. Buchanan R (1995) A dyer's garden: from plant to pot, growing dyes for natural fibres. Interweave, USA 112p
53. Chadramouli KV (1993) The color of our lives. PPST Foundation, Chennai, India 79p
54. Deveouglu O (2013) Marmara University, Natural Dye Researcher. http://mimoza.marmara. edu.tr/~ozan.deveoglu/A13.pdf. Accessed 20 Dec 2013
55. Grierson S, Duff DG, Sinclair RS (1985) Natural dyes of the Scottish highlands. Text Hist 16 (1):23–43
56. Gulrajani ML, Gupta D (eds) (1992) Natural dyes and their application to textiles. Department of Textile Technology, Indian Institute of Technology, New Delhi
57. Hill DJ (1997) Is there a future for natural dyes? Rev Prog Color 27:18–25
58. Siva R (2007) Status of natural dyes and dye-yielding plants in India. Curr Sci 92(7):916–925
59. Sharma KK, Pareek PK, Raja ASM, Temani P, Kumar Ajay, Shakyawar DB, Sharma MC (2013) Extraction of natural dye from *Kigelia pinnata* and its application on pashmina (cashmere) fabric. Res J Text Apparel 17(2):28–32
60. Gulrajani ML (1999) Present status of natural dyes. In: Book of papers of the convention on natural dyes, Department of Textile Technology, IIT Delhi, New Delhi, 9–11 Dec 1999
61. Nayak AKJR (2007) Natural dyeing and synthetic dyeing: a comparative cost analysis. Paper presented in stakeholders' workshop, ICEF Project on promotion of natural dyes in textile industries for environmental improvement and sustainable livelihood, Mumbai, 26 May 2007
62. Ince NH, Ziylan A (2015). Single and hybrid applications of ultrasound for decolorization and degradation of textile dye residuals in water. In: Sharma SK (ed) Green chemistry for dyes removal from wastewater: research trends and applications, pp 261–293
63. Heumann S, Eberl A, Pobeheim H, Liebminger S, Fischer-Colbrie G, Almansa E, Gübitz GM (2006) New model substrates for enzymes hydrolysing polyethyleneterephthalate and polyamide fibres. J Biochem Biophys Methods 69(1):89–99
64. Huang W, Leonas KK (2000) Evaluating a one-bath process for imparting antimicrobial activity and repellency to nonwoven surgical gown fabrics. Text Res J 70(9):774–782
65. Seong HS, Kim JP, Ko SW (1999) Preparing chito-oligosaccharides as antimicrobial agents for cotton. Text Res J 69(7):483–488
66. Chuang TH, Wu PL (2007) Cytotoxic 5-Alkylresorcinol metabolites from the leaves of *Grevillea robusta*. J Nat Prod 70(2):319–323
67. Han X, Shen T, Lou H (2007) Dietary phenols and their biological significance. Int J Mol Sci 8:950–988
68. Min BR, Pinchak WE, Merkel R, Walker S, Tomita G, Anderson RC (2008) Comparative antimicrobial activity of tannin extracts from perennial plants on mastitis pathogens. Sci Res Essay 2:66–73
69. Datta S, Uddin MA, Afreen KS, Akter S, Bandyopadhyay A (2013) Assessment of antimicrobial effectiveness of natural dyed fabrics. Bangladesh J Sci Ind Res 48(3):179–184
70. Gupta D, Khare SK, Laha A (2004) Antimicrobial properties of natural dyes against gram-negative bacteria. Color Technol 120(4):167–171
71. Prabhu KH, Teli MD (2011) Eco-dyeing using *Tamarindus indica* L. seed coat tannin as a natural mordant for textiles with antibacterial activity [Online]. J Saudi Chem Soc. doi:10. 1016/j.jscs.2011.10.014
72. Biradar YS, Jagatap S, Khandelwal KR, Singhania SS (2008) Exploring of antimicrobial activity of Triphala Mashian Ayurvedic formulation. eCAM 5(1):107–113
73. Sathianarayanan MP, Bhat NV, Kokate SS, Walunj VE (2010) Antibacterial finish for cotton fabric from herbal products. Indian J Fibre Text Res 35(1):50
74. Thilagavathi G, Kannaian T (2010) Combined antimicrobial and aroma finishing treatment for cotton, using microencapsulated geranium (Pelargonium graveolensL'Herit, ex Ait.) leaves extract. Indian J Nat Prod Resource 1(3):348–352

75. El-Tahlawy KF, El-Bendary MA, Elhendawy AG, Hudson SM (2005) The antimicrobial activity of cotton fabrics treated with different crosslinking agents and chitosan. Carbohydr Polym 60:421–430
76. Lim SH, Hudson SM (2003) Review of chitosan and its derivatives as antimicrobial agents and their uses as textile chemicals. J Macromol Sci Polym Rev 43:223–269
77. Zhang ZT, Chen L, Ji JM, Huang YL, Chen DH (2003) Antibacterial properties of cotton fabrics treated with Chitosan. Text Res J 73:1103–1106
78. Chattopadhyay SN, Pan NC, Roy AK, Saxena S, Khan BA (2013) Development of natural dyed jute fabric with improved color yield and UV protection characteristics. J Text Inst 104 (8):808–818
79. Katarzyna SP, Kowalinski J (2008) Light fastness properties and UV protection factor of naturally dyed linen, hemp and silk. In: Proceedings, FlaxBast: international conference on flax and other bast plants, Saskatoon, Canada, 21–23 July 2008
80. Sarkar AK (2004) An evaluation of UV protection imparted by cotton fabrics dyed with natural colorants. BMC Dermatol. doi:10.1186/1471-5945-4-15
81. Grifoni D, Zipoli G, Albanese L, Sabatini F (2011) The role of natural dyes in the UV protection of fabrics made of vegetable fibres. Dyes Pigm 91:279–285
82. Saxena S, Basak S, Mahangade RR (2013) Eco-friendly durable ultraviolet protective finishing of cotton textiles using pomegranate rinds. Paper presented at the international conference on advances in fibers, finishes, technical textiles and nonwovens, AATCC, Mumbai 1–2 Oct 2013
83. Specos MM, Garcia JJ, Tornesello J, Marino P, Della Vecchia M, Tesoriero MD, Hermida LG (2010) Microencapsulated citronella oil for mosquito repellent finishing of cotton textiles. Trans R Soc Trop Med Hyg 2010(104):653–658
84. Bhatt L, Kale R (2015) Development of mosquito repellent textiles using chrysanthemum oil nano emulsion. Int J Fashion Des Technol Educ 5(3):15–22
85. Basak S, Saxena S, Chattopadhyay SK, Narkar R, Mahangade R (2015) Banana pseudostem sap: a waste plant resource for making thermally stable cellulosic substrate. J Ind Text: 1528083715591580 (June 24)
86. Wang H, Lewis DM (2002) Chemical modification of cotton to improve fibre dyeability. Color Technol 118:159–168
87. Hashem MM (2007) An approach towards a single pretreatment recipe for different types of cotton. Fibres Text Eastern Europe 15(2) (61): 85–92
88. Fatma AM, El-Alfy EA (2013) Improving Dyebility of Cotton Fabric via Grafting with DimethylaminoEthylmethacrylate. J Appl Sci Res 9(1):178–183
89. Wu TS, Chen KM (1993) New cationic agents for improving the dyeability of cotton fibers. J Soc Dyers Colour 109:153–157
90. Burkinshaw SM, Lei XP, Lewis DM (1989) Modification of cotton to improve its dyeability. Part 1-pretreating cotton with reactive polyamide-epichloro-hydrin resin. J Soc Dyers Colour 105:391–398
91. Blackburn RS, Burkinshaw SM (2003) Treatment of cotton with cationic, nucleophilic polymers to enable reactive dyeing at neutral pH without electrolyte addition. J Appl Polym Sci 89:1026–1031
92. Burkinshaw SM, Mignanelli M, Froehling PE, Bride MJ (2000) The use of dendrimers to modify the dyeing behavior of reactive dyes on cotton. Dyes Pigm 2000(47):259–267
93. Zhang F, Chen Y, Lin H, Lu Y (2007) Synthesis of an amino-terminated hyperbranched polymer and its application in reactive dyeing on cotton as a salt-free dyeing auxiliary. Coloration Tec 123:351–357
94. Gentile DB (2009) A thesis titled reduced salt usage in dyeing of 100 % cotton fabric, by School of Fashion & Textiles College of Design & Social Context, RMIT University
95. Lim SH, Hudson SM (2004) Application of a fiber-reactive chitosan derivative to cotton fabric as a zero-salt dyeing auxiliary. Color Technol 2004(120):108–113
96. Prabu HG, Sundrarajan M (2002) Effect of the bio-salt trisodium citrate in the dyeing of cotton. Color Technol 118:131–134

97. Khatri A, Padhyea R, Whitea M (2012) The use of tri sodium nitrilo triacetate in the pad–steam dyeing of cotton with reactive dyes. Color Technol 129:76–81
98. Ahmed NSE (2005) The use of sodium edate in the dyeing of cotton with reactive dyes. Dyes Pigments 65(3):221–225
99. Yu B, Wang WM, Cai ZS (2014) Application of sodium oxalate in the dyeing of cotton fabric with reactive red 3BS. J Text I 105(3):321–326
100. Guan Y, Zheng Qing-kang, Mao Ya-hong, Gui Ming-sheng, Hong-bin Fu (2007) Application of polycarboxylic acid sodium salt in the dyeing of cotton fabric with reactive dyes. J Appl Polym Sci 105(2):726–732
101. Gupta S (1996) Recent advances in wet chemicals processing machinery. Indian J Fibre Text Res 21:57–63
102. Ali SS, Khatri Z, Brohi KM (2012) Econtrol dyeing process: an ecological and economical approach. In: Aslam UM, Khanji H (eds) Energy, environment and sustainable development. Springer, Vienna, pp 291–297
103. Lu X, Liu L, L R, Chen Jihua (2010) Textile wastewater reuse as an alternative water source for dyeing and finishing processes: a case study. Desalination 258(1–3):229–232
104. Lanza KM (2016) An alternative for textile wastewater. Technical paper, GE water and process technologies. Available from https://www.gewater.com/kcpguest/document-library. do. Accessed 13 May 2016
105. Dascalu T, Acosta-Ortiz SE, Ortiz-Morales M (2000) Removal of indigo color by laser beam-denim interaction. Opt Laser Eng 34:179–189
106. Kan CW (2015) Washing techniques for denim jeans. In: Paul R (ed) Denim: manufacture, finishing and applications. Elsevier, Cambridge
107. Ondogan Z, Pamuk O, Ondogan EN, Ozguney A (2005) Improving the appearance of all textile products from clothing to home textile using the laser technology. Opt Laser Technol 37(8):631–637
108. Van der Kraan M (2005) Process and equipment development for textile dyeing in supercritical carbon dioxide. Delft University of Technology, TU Delft
109. Wen H, Dai JJ (2007) Dyeing of polylactide fibers in supercritical carbon dioxide. J Appl Polym Sci 105(4):1903–1907
110. Russell M (2014) Waterless dyeing gets mixed industry reaction. http://www.just-style.com/analysis/waterless-dyeing-garners-mixed-industry-reaction_id120351.aspx. Accessed 22 April 2016
111. Beck KR, Lynn GM (1997) Extraction of cotton impurities: supercritical CO_2 vs. Soxhlet/TCE. Text Chem Colorist 29(8):88–70
112. Cai Z, Qui Y, Zhang C, Hwang YJ, McCord M (2003) Effect of atmospheric plasma treatment on desizing of PVA on cotton. Text Res J 73(8):670–674
113. Kan CW, Lam CF, Chan CK, Ng SP (2014) Using atmospheric pressure plasma treatment for treating grey cotton fabric. Carbohydr Polym 102(15):167–173
114. Sun D, Stylios GK (2004) Effect of low temperature plasma treatment on the scouring and dyeing of natural fabrics. Text Res J 74(9):751–756
115. Long JJ, Wang HW, Lu TQ, Tang RC, Zhu YW (2008) Application of low-pressure plasma pretreatment in silk fabric degumming process. Plasma Chem Plasma Process 28(6):701–713
116. Mehmood T, Kaynak A, Dai XJ, Kouzani A, Magniez K, de Celis DR, Hurren CJ, du Plessis J (2014) Study of oxygen plasma pre-treatment of polyester fabric for improved polypyrrole adhesion. Mater Chem Phys 143(2):668–75
117. Kan CW, Yuen CWM (2006) Low temperature plasma treatment for wool fabric. Text Res J 76(4):309–314
118. Teli MD, Samanta KK, Pandit P, Basak S, Chattopadhyay SK (2014) Low temperature dyeing of silk using atmospheric pressure plasma treatment. Indian J Nat Fibres 1(1):1–7
119. Man WS, Kan CW, Ng SP (2014) The use of atmospheric pressure plasma treatment on enhancing the pigment application to cotton fabric. Vacuum 99:7–11

120. Kan CW, Yuen C, Tsoi W (2011) Using atmospheric pressure plasma for enhancing the deposition of printing paste on cotton fabric for digital ink-jet printing. Cellulose 18(3):827–839
121. Hartwig H (2002) Plasma treatment of textile fibers. Pure Appl Chem 74(3):423–427
122. Li S, Jinjin D (2007) Improvement of hydrophobic properties of silk and cotton by hexa fluoro propene plasma treatment. Appl Surf Sci 253(11):5051–5055
123. Samanta KK, Jassal M, Agrawal AK (2010) Atmospheric pressure plasma polymerization of 1, 3-butadiene for hydrophobic finishing of textile substrates. J Phys Conf Ser 208:012098
124. David JM, Zachary RD, Robert JS, Erik H, Jeong-Hoon K, Jung-Gu K, Seong HK (2013) Atmospheric rf plasma deposition of superhydrophobic coatings using tetramethylsilane precursor. Surf Coat Tech 234(15):14–20
125. Diener E (2006) Plasma surface technology. http://www.plasma-us.com/files/diener_web_en.pdf. Accessed 28 Aug 2014
126. Dow corning corporation (2007) Dow corning plasma solutions' application note (online). http://www.dowcorning.com/content/publishedlit/01-3137-01.pdf. Accessed 28 Aug 2014
127. Sigma Technologies (2006) Sigma technologies atmospheric plasma treaters for high-speed web applications (online) http://sigmalabs.squarespace.com/storage/publications-and-resources/SIGMA%20APT%20Brochure.pdf. Accessed 28 Aug 2014
128. Anon (2016) http://www.textileexcellence.com/news/details/1092/oeko-tex-launches-new-eco-passport-certification-for-sustainable-textile-chemicals#sthash.j5JwXZJG.dpuf. Accessed 18 April 2016
129. Acxys http://www.acxys.com/textile.html. Accessed 28 Aug 2014
130. Airoli, Plasma (Brochure) (online) http://www.arioli.biz/images/brochure/catalogo_plasma.pdf. Accessed 28 Aug 2014
131. Apjet http://www.apjet.com/. Accessed 28 Aug 2014
132. Europlasma (2013) Press release Europlasma launches pfoa- and pfos-free nanocoatings for technical textiles under brand name "nanofics" (online) http://www.europlasma.be/uploads/content/files/PressRelease20130611Techtextil.pdf. Accessed 28 Aug 2014
133. Grinp http://www.grinp.com/. Accessed 28 Apr 2016
134. Plasmatreat http://www.plasmatreat.com/. Accessed 28 Aug 2014
135. Softal (2014) http://www.softal.de/de/loesungen/oberflaechenbehandlung-mit-der-aldyne-technologie/ Accessed 28.8.2014
136. Ströhle J, Schramek G (2016) Textiles achieve ecological footprint, new opportunities for China. http://www.benningergroup.com/uploads/media/Carbon_Footprint_China_EN.pdf. Accessed 28 Mar 2016
137. Vitoplasma (2014) http://www.vitoplasma.com/en/plasmazone. Accessed 28 Aug 2014

Anthraquinone-based Natural Colourants from Insects

Shahid-ul-Islam and F. Mohammad

Abstract Increasing concerns about the rising pollution in the recent years has resulted in creation of more awareness about the use of natural products in textile applications and has led to strong consumer demand for the use of 'green' products based on renewable materials. Subsequently, in textile industries, scientists and polymer chemists have focussed their momentum towards the development of substitute to synthetic dyeing and finishing agents. It is only in the past few decades or so the joint efforts by scientists, academicians and R&D organizations have dramatically increased the market for colourants extracted from plants, minerals and insects. Insect-derived colourants, among different dye sources, have opened new interesting fields for textile and polymer researchers. This chapter first highlights sources, classification and finally discusses the production of natural colourants from well-known dye-bearing insects for use in textile dyeing and finishing.

Keywords Dyes · Insects · Kermes · Lac · Cochineal

1 Introduction

In primitive times, crude extracts from seeds, flowers, stems, barks, berries and leaves from various plant species have been used for coloration of textiles and decoration of human caves and dwellings [4, 29, 30]. After the discovery of weaving at about 5000 B.C., it is well documented that Egyptians started dyeing by 3000 B.C. Recorded history has shown that the nails of Egyptian Mummies were dyed with *Lawsonia inermis*. Ancient Egyptians were using naturally occurring coloured minerals for various colours required. They also used several vegetable dyes: madder, safflower and alkanet for red; indigo for blue; and the bark of the pomegranate tree for yellow. Alizarin, a red pigment extracted from madder, has been detected on red fabrics found in Tutankhamun's tomb in Egypt [47].

Shahid-ul-Islam (✉) · F. Mohammad
Department of Chemistry, Jamia Millia Islamia, New Delhi 110025, India
e-mail: shads.jmi@gmail.com

© Springer Science+Business Media Singapore 2017
S.S. Muthu (ed.), *Textiles and Clothing Sustainability*,
Textile Science and Clothing Technology,
DOI 10.1007/978-981-10-2185-5_3

With the discovery of first violet synthetic dye 'Mauveine' by William Henry Perkin, in 1856, the use of natural dyes in textile coloration industries decreased to a large extend [1, 3, 7, 26]. Synthetic dyes received faster global acceptability due to wide range of applications in various fields, and reproducibility in shades and overall cost factor combined with inherent problems encountered with natural dye applications [16, 36]. Synthetic dyestuff, in particular azo dyes, may cause skin allergies, waste products and have come under severe criticism for their high environmental pollution, at the stage of manufacturing as well as application. Owing to this, strict ecological and economic restrictions over the applied chemicals, including bans on some synthetic dyes (e.g. azo and benzidine dyes), have been imposed by many countries including Germany, European Union, USA and India [60]. As a result, there has been great motivation in the reintroduction of natural dyes in textile coloration [45, 59]. Although development of research in the area of natural dyes since its inception started mainly in Egypt, India, Iran, South Korea, Pakistan, Thailand and Bangladesh, the significant revival of interest in natural dyes and colourants in various application disciplines was demonstrated by the overwhelming response obtained from several traditional dyers/crafts persons, scientists, textile artists, professionals from different branches of textile industry, government representatives from around the globe at a series of recent International Symposia/Workshops on Natural Dyes [25, 54, 60, 70].

In addition to the development of uncommon, harmonizing sober and elegant shades on different textile materials, natural colourants from plant sources have been recently discovered as novel agents in imparting multifunctional properties such as antimicrobial [62], anti-feedant [34], deodorizing [27] and UV protection properties [58].

A number of limitations associated with colourants from natural materials are highlighted by Glover [19] and are described below.

- Inadequate availability of natural dyes for global consumption
- Poor yield of colouring matters
- Difficulties in reproducibility
- Limited shade range
- Lack of standard procedure for extraction and dyeing processes
- Unsuitability for synthetic fibres
- Low dye exhaustion and poor fastness properties
- High costs of extraction, mordanting and dyeing
- Lack of scientific knowledge on compatibility and chemistry of natural dyes
- Use of metallic salts mordants.

To overcome the above-mentioned limitations over the past few years, scientists have applied apart from conventional exhaustion methods [5] a variety of non-conventional methods such as ultrasonic [32], microwave [50], high-temperature high-pressure (HTHP) dyeing [8], contact dyeing [31] and other radiation treatments including gamma and UV [2, 3, 7] to increase extraction as well dyeing and fastness properties of textile materials. The aim of the present

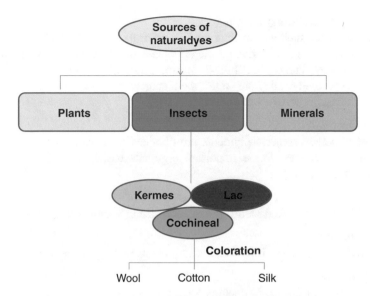

Fig. 1 Diagrammatic view of different sources of natural colourants

chapter was to highlight sources, classification and finally to discuss the production of natural colourants from insect sources for use in textile dyeing and finishing (Fig. 1).

2 Classification of Natural Dyes

Natural dyes are colourants obtained from plants (flowers, seeds, leaves, barks, trunks and roots), insects (lac, cochineal and kermes), animals (molluscs) or mineral substances without chemical processing and have been used in coloration of food, leather, silk, wool and cotton since prehistoric times. Natural dyes have been classified on the basis of sources, application and chemical structures [21, 42, 57].

2.1 Classification on the Basis of Sources

Natural dyes can be classified into three categories on the basis of their origin:

Vegetable Origin
Majority of the natural dyes are obtained from plant sources. The colouring matters derived from different parts of plants such as flowers, fruits, seeds, leaves, barks, trunks and roots fall in this category [63].

Insect/Animal (Molluscs) Origin

Natural substances such as carminic acid (cochineal), kermesic acid (kermes), and laccaic acid (lac dye) and Tyrian purple are obtained from either exudation or dried bodies of insects, namely Cochineal, Kermes, *Laccifer lacca/Kerria lacca* and molluscs, respectively. They are well-known sources of natural dyes and have been used for dyeing purposes from ancient times [9].

Mineral Origin

The most important mineral pigments are obtained from various inorganic metal salts and metal oxides. Some important pigments based on mineral origin are enlisted as follows [57]:

Red Pigments

Cinnabar (Vermilion), Red Ochre (Geru), Red lead (Sindur/Minium), Realgar.

Yellow Pigments

Yellow Ochre (Ram Raj), Raw Sienna, Orpiment, Litharge (Massicot).

Green Pigments

Terre-Verte (Green Earth), Malachite, Vedgiris.

Blue Pigments

Ultramarine Blue, Azurite.

White Pigments

Chalk (White Lime), White lead, Zinc White.

Black Pigments

Charcoal Black, Lamp Black, Ivory Black, Bone Black, Graphite, Black Chalk, Terre-noire (Black Earth).

2.2 Classification on the Basis of Method of Application

According to this classification, the colourants which dye the fibres/fabrics directly are classified as **substantive** dyes, such as Indigo, Turmeric and Orchil. **Adjective** dyes are those which dye the fibres/fabrics mordanted with a metallic salt or with the addition of a metallic salt to the dye bath, examples of such dyes are Cochineal, Fustic, Logwood and Madder [20, 57]. Monogenetic and polygenetic are two different classes of dyes. **Monogenetic** dyes (annatto) produce only one colour irrespective of the mordant present on the fibre or applied along with the dye; **polygenetic** dyes produce different colours according to the mordant employed and examples of this class are logwood, kamala, lac and cochineal [56].

Based on method of application, natural dyes can be classified into following classes:

Mordant Dyes

Mordant dyes are those dyes which have higher affinity for the mordanted fibres/fabrics. This classical definition of mordant dyes has been extended to cover all those dyes which form a coordination complex with the metal mordant [56]. The complex may be formed by first applying the mordant (pre-mordanting) or by simultaneous application of the mordant and the dye (simultaneous mordanting) or by after treatment of the dyed material with mordant (post-mordanting) [40]. Most of these dyes yield different shades or colours with different mordants. The final colour of a mordant dye depends upon the metal salt employed [69]. Mordants also improve light, wash and rub fastness properties [70].

Vat Dyes

Vat dyes are a class of colourants important for coloration and printing on cotton- and cellulosed-based textiles. Vat dyes are insoluble in their coloured form; how-ever, it can undergo reduction into soluble colourless (leuco) form which has an affinity for fibre or textile to be dyed. Re-oxidation of the vat dyes converts them again into insoluble form with retention of original colour [21]. The vat dyes derived their name from the 'vat' which was at one time used for reducing the dye and converting it into soluble form. Only three natural dyes belong to vat dyes: indigo, woad and tyrian purple [9].

Indigo   Tyrian Purple

Direct Dyes

Direct dyes are water-soluble organic molecules which can be applied as such to the fibres or textiles, since they have affinity and are taken up directly [9]. These dyes are used for colouring of cotton, leather, wool, silk and nylon. Direct dyeing is normally carried out in a neutral or nearly alkaline dyebath to yield bright colours [21]. However, due to the nature of chemical interaction, their fastness to washing is poor, although this can be improved by special after-treatment. Some prominent examples of direct natural dyes are turmeric and annatto [28].

$R_1, R_2 = OCH_3$ - Curcumin

$R_1 = OCH_3, R_2 = H$ - Demethoxycurcumin

$R_1, R_2 = H$ -Bisdemethoxycurcumin

Acid Dyes

Acid dyes are water-soluble anionic dyes to the polyamide fibres such as wool, silk and nylon. These dyes are applied in acidic medium, and they have either sulphonic acid or carboxylic acid groups in the dye molecules [10, 63]. The dye–fibre interaction is attributed to the salt formation between cationic groups in fibre and anionic groups in dye structure [38, 56]. Saffron is one such natural dye obtained from *Crocus sativus* which has been classified as acid dye. This dye has two carboxylic acid groups [13]. These groups can form an electrovalent bond with the amino groups of wool and silk [46].

Gentiobiosyl (Gen)   Glucosyl (Glu)

Crocetin $R_1 = R_2 = H$ Crocin-1: $R_1 = R_2 = $ Gen

Crocin-2 $R_1 = $ Gen, $R_2 = $ Glu   Crocin-3: $R_1 = $ Gen, $R_2 = H$

Basic Dyes

Basic dyes are water-soluble cationic dyes and normally applied to acrylic fibres. These dyes are known to form coloured cations upon ionization which form an electrovalent bond with the carboxyl group of wool and silk [41, 64]. These dyes are applied from neutral to mild acidic condition. Berberine, a natural pigment, has been classified as a basic dye [9].

Berberine

2.3 Classification on the Basis of Chemical Structure

This is the most appropriate and widely used system of classification of natural dyes, because it readily identifies dyes belonging to a chemical group which has certain characteristic properties (Fig. 2). Different chemical classes of natural dyes are discussed below [9, 21].

3 Anthraquinones

This is the largest group of natural quinines and is building block of many natural dyes, particularly the important red dyes. They are obtained both from plants and from insects. These dyes form stable coordination complexes with metal salts or mordants, and the resultant metal-complex dyes have good light and wash fatness properties. Some important dye yielding plants belonging to this group are discussed below.

9, 10-Anthraquinone

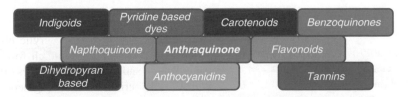

Fig. 2 Classification of natural colourants on the basis of chemical structures

3.1 Rheum emodi

Indian Rhubarb scientifically known as *Rheum emodi* has been used since time immemorial in Ayurvedic and Unani systems of medicine. It is found in temperate and subtropical regions of the world, Himalayas in India and China [39]. Indian rhubarb or emodi contains diverse chemical compound flavonoids including rutin, puerarin and genistein [53]. Besides its roots contain a large number of anthraquinone derivatives such as rhein, emodin, aloe emodin, chrysophanol and physcion. These occur free and as quinine, anthrone or dianthrone glycosides [35]. The isolated anthraquinone derivatives isolated from its rhizomes are highly active against a number of fungal pathogens including *Candida albicans*, *Cryptococcus neoformans*, *Trichophyton mentagrophytes* and *Aspergillus fumigatus*.

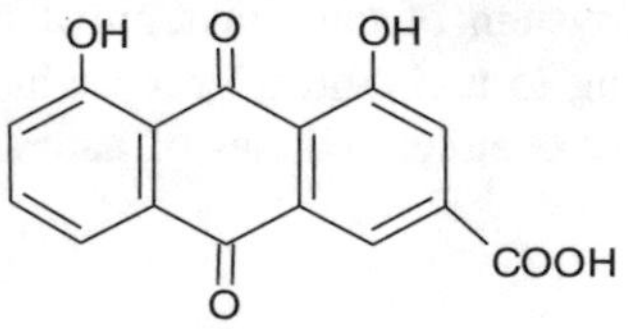

Rhein

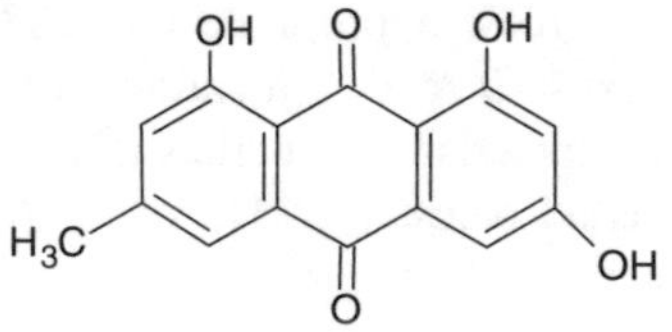

Emodin

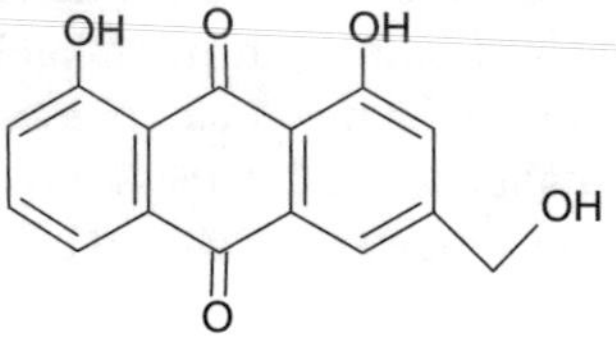

Aloe-emodin

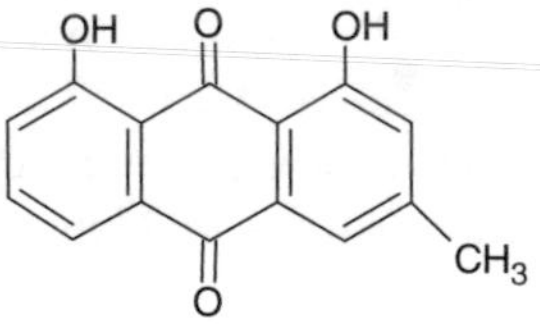

Chrysophaol

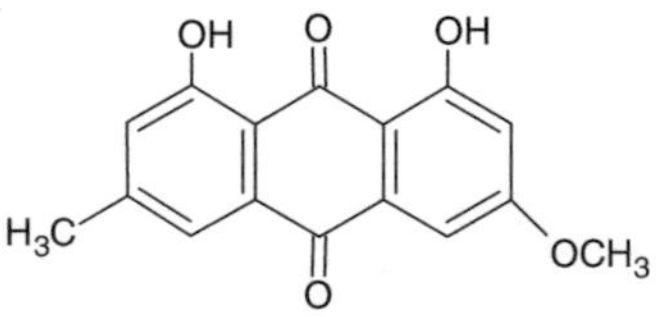

Physcion

The herb is also chiefly used in medicine as a purgative astringent tonic. The phytochemical present in madder roots has potential antimicrobial properties [39]. The colourants extracted from roots have been used in colouring food stuffs and wool, cotton and other textiles [17, 37, 49].

3.2 Rubia tinctorum *(Madder)*

Madder is a well-known dye plant native to Western and Central Asia [71]. Madder parts have been traditionally used to combat various infections including kidney and bladder stones. It is believed that the cultivation of madder sharply declined after the introduction of synthetic alizarin in second half of the nineteenth century [9, 53]. Madder dyes are hydroxyl anthraquinones, which are extracted from the ground root of the plants of Rubia species of family *Rubiaceae,* e.g. *Rubia tinctorum* [65].

Madder has been important source of red dye for textile coloration. The main colouring compounds isolated from madder roots are alizarin, purpurin, pseudopurpurin, rubiadin, munjistin and xanthopurpurin, of which alizarin is the major one [43, 61, 70].

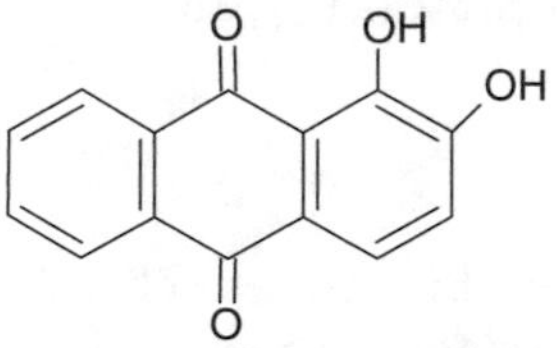

Alizarin

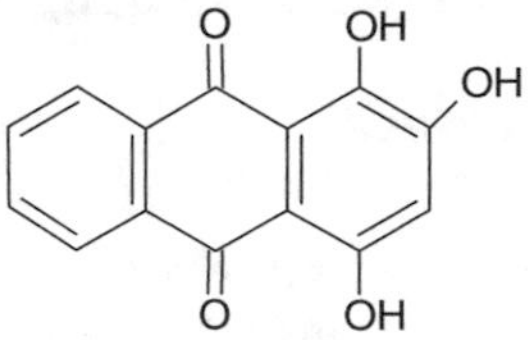

Purpurin

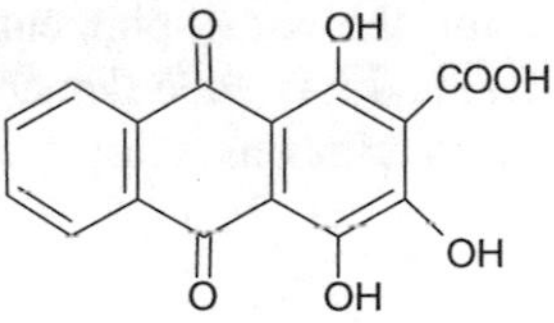

Pseudopurpurin

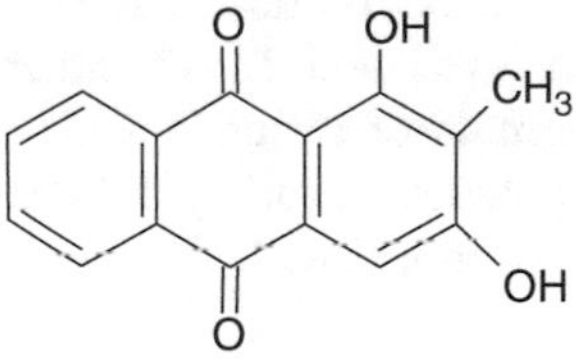

Rubiadin

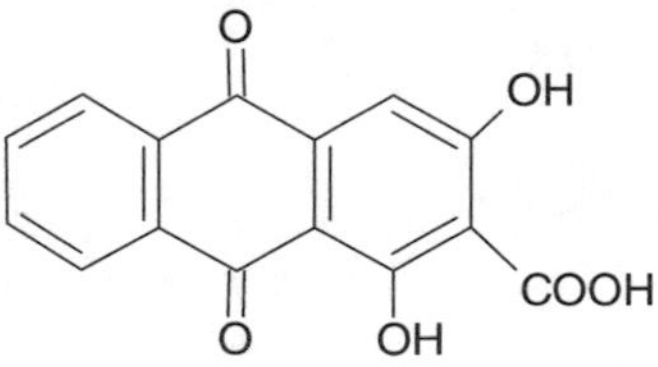

Munjistin

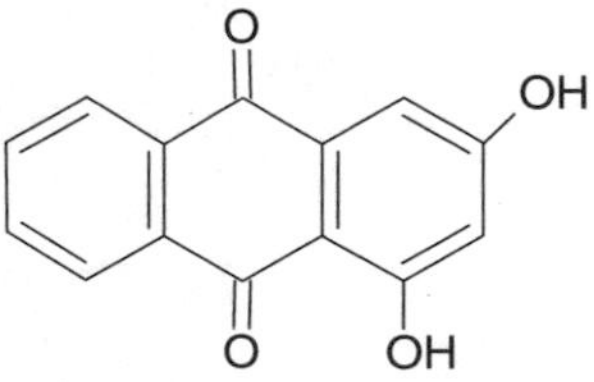

Xanthopurpurin

Investigations carried out so far have shown that alizarin is not present in Indian madder (*Rubia cordifolia*), which contains approximately 10 colourants having anthraquinoid structure [23]. The major constituents are purpurin (1,2,4-trihydroxyanthraquinone), (65–70 %), munjistin (1,3-dihydroxy-2-carboxyanthraquinone (10–12 %) and nordamncanthal 1,3-dihydroxy-2-formylanthraquinone (9–10 %) [24].

Madder is classified as a mordant dye and has been used to develop wide range of shades varying from red and brown on wool and cotton using different metal salt mordants such as ferrous sulphate, stannous chloride and aluminium sulphate [51]. The colouring compounds isolated from madder roots exhibit antibacterial and antifungal properties; therefore, it is used for simultaneous coloration as well as antimicrobial modifications [52, 61].

4 Insect Dyes Belonging to Anthraquinone Group

4.1 Cochineal

Cochineal is a natural colourant extracted from the dried bodies of female bodies of a scale-insect, *Dactylopius coccus*. The main hosts for these female insects are the aerial parts of prickly pear Opuntia and Wild *cacti* found in Mexico in Central America and Brazil in South America [48]. Carminic acid is the main red pigment which has been used in food coloration, pharmaceutical applications and in the dyeing of wool, cotton and silk. *D. confusus*, *D. indicus Green*, *D. tomentosus*, *Anthracoccus uvae-ursi* L. are other sources for the same pigment [66]. The molecular formula of carminic acid is $C_{22}H_{20}O_{13}$ and has wide demand in textile dyeing as it produces a variety of hues ranging from crimsons, scarlets and pinks in the presence of mordants [22].

Carminic acid

4.2   Kermes

Kermes is another natural dye basically consists of the dried insect of *Kermes* (*Coccus ilici/Kermes vermillio*). These insects feed on evergreen Oak (*Quercus coccifera*) and are collected in the month of June and are being killed by vapours of acetic acid [28]. Kermes has been used since ancient times in England, Morocco, Scotland, South France, Spain and Turkey being used for dyeing wool and leather [48].

Kermesic acid which is aglycone of carminic acid is the main colouring component present in kermes. It is slightly soluble in cold water, soluble in hot water giving an yellowish-red solution [18]. Kermes gives different shades on wool depending on the type of mordant used [67].

Kermesic acid

4.3   Lac Dye

Lac is natural reddish dyes used since prehistoric times and is probably the most ancient of insect dyes. This colouring component is extracted from stick lac which is a secretion of insect *Laccifer lacca/Kerria lacca* [11, 21]. The insect *Laccifer lacc* is found growing on the twigs of certain rain trees such as *Samanea saman* (Jacq.) Merr. (*Pithecolobium saman*, Mimosaceae native to India and Southeast Asia. Lac natural dye is composed of mixture of five closely related laccaic acids such as laccaic acids A, B, C, D and E all having anthraquinoid structure with two carboxylic acid groups [12, 72].

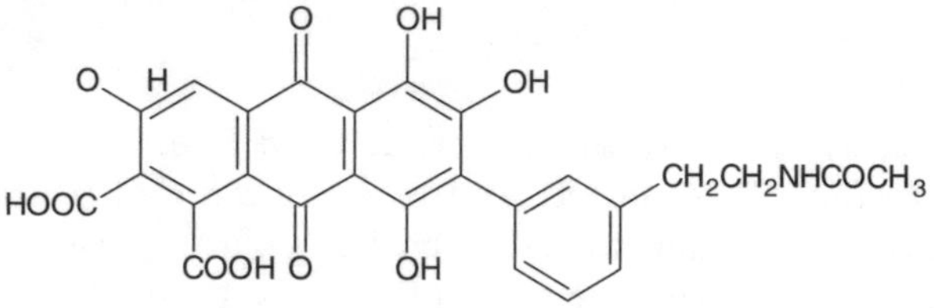

Laccaic acid A

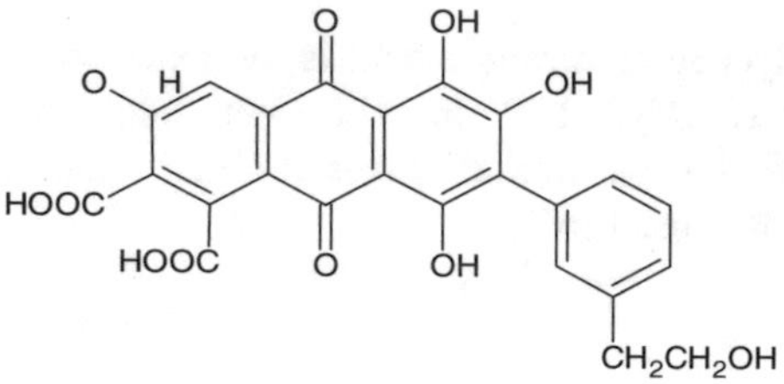

Laccaic acid B

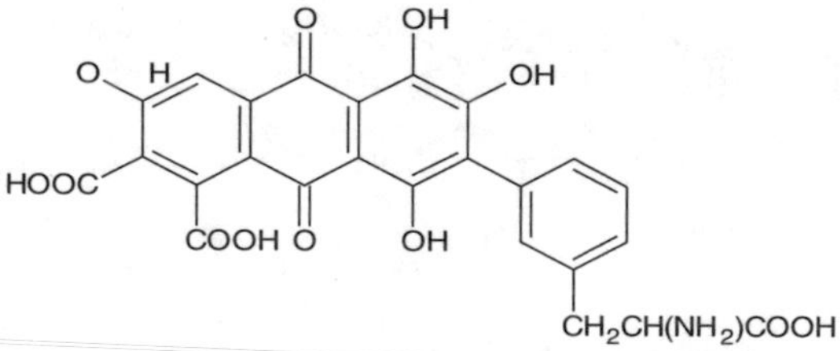

Laccaic acid C

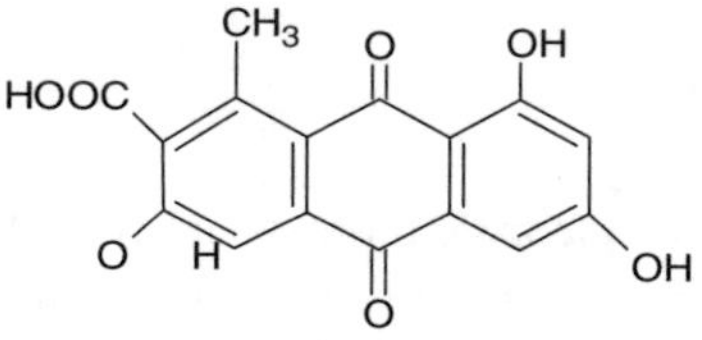

Laccaic acid D

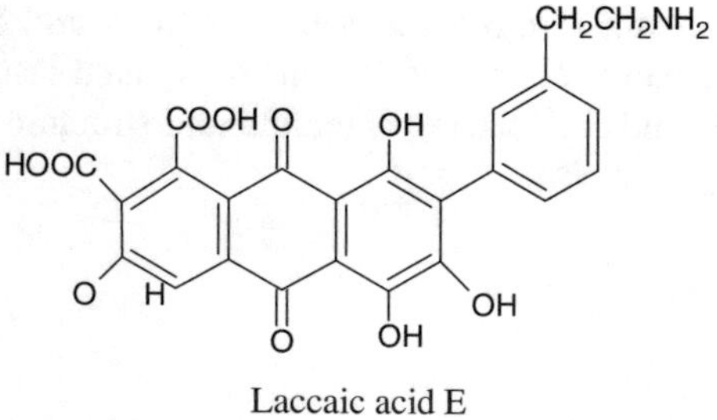

Laccaic acid E

The dye extracted from lac soluble in a number of solvents including methyl alcohol, acetone and acetic acid [68]. With two carboxyl acid groups in its structure, lac is considered as acidic dye. Lac dissolved at acidic pH has been used in wool and silk dyeing [33]. More recently, research investigations have been carried to study the thermodynamics and kinetic of lac dyeing onto silk and cotton. Janhom et al. used poly(ethyleneimine) (PEI) to increase the lac dye uptake by cotton. Kamel et al. [33], in a research experiment, examined the influence of conventional and ultrasonic techniques on both dye extraction and dyeing properties of lac onto wool. Different dyebath parameters such as pH, salt concentration, ultrasonic power, time and temperature were studied, and it was found that ultrasound dyeing is the effective way to obtain deep shades with good fastness properties than conventional method. Kamel et al. [32] also proved that ultrasound assisted dyeing of cationized cotton yield better results than conventional dyeing. Chairat et al. [14] evaluated the adsorption and kinetic studies of lac dyeing of silk and found that Langmuir and Freundlich isotherms were best fitted with the experimental data. They also found that the dyeing time of 60 min, pH 3, material to liquor ratio of 1:100 and an initial dye concentration of 450 mg/L are the optimum conditions for efficient dyeing of silk. In a subsequent experiment, Chairat et al. [15] investigated also adsorption kinetic studies of lac dyeing on cotton and discovered that pseudosecond-order kinetic model with activation energy of 42.4 kJ/mol gives the best fit. Likewise, Rattanaphani et al. [55] studied the adsorption and kinetic of lac dyeing onto cotton at pH 3.0, a material to liquor ratio of 1:100 and a contact time of 3 h previously pretreated with chitosan. It was observed that chitosan improved the dyeing performance of lac dye by increasing its uptake from dyebath. Langmuir isotherm was found to fit with the experimental results giving an enthalpy change ($\Delta H°$) of -17.43 kJ. Similarly, silk surface was modified using oxygen and argon plasma treatments by Boonla and Saikrasun [6] to investigate the adsorption and kinetic of lac dye. Argon treatment gave the best results than oxygen by increasing dye uptake and fastness properties. It also produces beautiful shades on different textile materials [44].

5 Conclusion and Future Perspectives

In general, it has been demonstrated that using natural colourants from different plant and insect sources is becoming a promising alternative to replace petrochemical derived dyes. Although dyes from natural materials have low dye exhaustion and fastness properties, the use of metal mordants, radiation modifications, after treatments, plasma and chitosan applications has the ability to overcome these problems. The literature indicates that significant advancements have been witnessed in dyeing of wool, cotton, silk and others with insect-derived dyes recently, but still there are number of challenges or gaps in our knowledge that must be overcome before insect colourants can practically be applied commercially instead of only at laboratory scale. The required scientific studies and systematic reports are severely missing as only three insect dyes have been explored until now. This field of research has more room for the improvement as there are a vast number of insects which could serve as source for eco-friendly natural dyes. More and more research should be carried out to re-establish the traditional dyeing techniques, and innovative methods should be explored for maximum extraction of colouring matter. In addition, future research should mainly focus on modern-day technologies in order to improve colour fastness or dyeing characteristics of dyed fabrics.

References

1. Adeel S, Ali S, Bhatti IA, Zsila F (2009) Dyeing of cotton fabric using pomegranate (*Punica granatum*) aqueous extract. Asian J Chem 21:3493–3499
2. Adeel S, Bhatti IA, Kausar A, Osman E (2012) Influence of UV radiations on the extraction and dyeing of cotton fabric with *Curcuma longa* L. Indian J Fibre Text Res 37:87–90
3. Adeel S, Fazal-ur-Rehman, Hanif R, Zuber M, Ehsan-ul-Haq, Muneer M (2014) Ecofriendly dyeing of UV-irradiated cotton using extracts of acacia nilotica bark (Kikar) as source of Quercetin. Asian J Chem 26:830–834
4. Ajmal M, Adeel S, Azeem M, Zuber M, Akhtar N, Iqbal N (2014) Modulation of pomegranate peel colourant characteristics for textile dyeing using high energy radiations. Ind Crops Prod 58:188–193
5. Ali S, Hussain T, Nawaz R (2009) Optimization of alkaline extraction of natural dye from Henna leaves and its dyeing on cotton by exhaust method. J Clean Prod 17:61–66
6. Boonla K, Saikrasun S (2012) Influence of silk surface modification via plasma treatments on adsorption kinetics of lac dyeing on silk. Text Res J 0040517512458344
7. Batool F, Adeel S, Azeem M, Ahmad Khan A, Ahmad Bhatti I, Ghaffar A, Iqbal N (2013) Gamma radiations induced improvement in dyeing properties and colorfastness of cotton fabrics dyed with chicken gizzard leaves extracts. Rad Phys Chem 89:33–37
8. Bhattacharya NL (2002) Dyeing of cotton and polyester fiber with pomegranate rind, catechu, nova red and turmeric. Asian Text J 11:70–74
9. Bhattacharyya N (2010) Natural dyes for textiles and their eco-friendly applications. IAFL Publications, New Delhi, India
10. Bliss A (1981) A handbook of dyes from natural materials. Charles Scribner's Sons, New York
11. Burwood R, Read G, Schofield K, Wright D (1965) The pigments of stick lac. Part I. Isolation and preliminary examination. J Chem Soc 6067–6073
12. Burwood R, Read G, Schofield K, Wright DE (1967) The pigments of stick lac: Part II. The structure of laccaic acid A. J Chem Soc 9:842–851
13. Caballero-Ortega H, Pereda-Miranda R, Abdullaev FI (2007) HPLC quantification of major active components from 11 different saffron (*Crocus sativus* L.) sources. Food Chem 100:1126–1131
14. Chairat M, Rattanaphani S, Bremner JB, Rattanaphani V (2005) An adsorption and kinetic study of lac dyeing on silk. Dyes Pigm 64:231–241
15. Chairat M, Rattanaphani S, Bremner JB, Rattanaphani V (2008) Adsorption kinetic study of lac dyeing on cotton. Dyes Pigm 76:435–439
16. Chattopadhyay SN, Pan NC, Roy AK, Saxena S, Khan A (2013) Development of natural dyed jute fabric with improved colour yield and UV protection characteristics. J Text Inst 104:808–818
17. Das D, Maulik SR, Bhattacharya SC (2008) Colouration of wool and silk with *Rheum emodi*. Indian J Fibre Text Res 33:163–170
18. Gadgil DD, Rama Rao AV, Venkatarman K (1968) Structure of kermesic acid. Tetrahedron Lett 9:2223–2227
19. Glover B (1998) Doing what comes naturally in the dyehouse. J Soc Dyers Colour 114:4–7
20. Gulrajani ML (2001) Present status of natural dyes. Indian J Fibre Text Res 26:191–201

21. Gulrajani ML, Gupta D (1992) Natural dyes and their application to textiles. Department of Textile Technology, IIT Delhi, New Delhi, India
22. Gupta D, Bhaumik S (2007) Antimicrobial treatments for textiles. Indian J Fibre Text Res 32:254–263
23. Gupta D, Kumari S, Gulrajani M (2001) Dyeing studies with hydroxyanthraquinones extracted from Indian madder. Part 1: dyeing of nylon with purpurin. Color Technol 117:328–332
24. Gupta D, Kumari S, Gulrajani M (2001) Dyeing studies with hydroxyanthraquinones extracted from Indian madder. Part 2: dyeing of nylon and polyester with nordamncanthal. Color Technol 117:333–336
25. Hill D (1997) Is there a future for natural dyes? Rev Prog Color 27:18–25
26. Holme I (2006) Sir William Henry Perkin: a review of his life, work and legacy. Color Technol 122:235–251
27. Hwang EK, Lee YH, Kim HD (2008) Dyeing, fastness, and deodorizing properties of cotton, silk, and wool fabrics dyed with gardenia, coffee sludge, *Cassia tora.* L., and pomegranate extracts. Fiber Polym 9:334–340
28. Islam S-u, Mohammad F (2015) Natural colorants in the presence of anchors so-called mordants as promising coloring and antimicrobial agents for textile materials. ACS Sustain Chem Eng 3:2361–2375
29. Islam S-u, Mohammad F (2014) Emerging green technologies and environment friendly products for sustainable textiles. In: Muthu SS (ed) Roadmap to sustainable textiles and clothing. Springer, Singapore, pp 63–82
30. Islam S, Shahid M, Mohammad F (2013) Perspectives for natural product based agents derived from industrial plants in textile applications—a review. J Clean Prod 57:2–18
31. Kadolph SJ (2008) A traditional craft experiencing new attention. Delta Kappa Gamma Bull 75:14–17
32. Kamel MM, El-Shishtawy RM, Youssef BM, Mashaly H (2007) Ultrasonic assisted dyeing. IV. Dyeing of cationised cotton with lac natural dye. Dyes Pigm 73:279–284
33. Kamel MM, El-Shishtawy RM, Yussef BM, Mashaly H (2005) Ultrasonic assisted dyeing. III. Dyeing of wool with lac as a natural dye. Dyes Pigm 65:103–110
34. Kato H, Hata T, Tsukada M (2004) Potentialities of natural dyestuffs as antifeedants against varied carpet beetle, *Anthrenus verbasci*. Jpn Agric Res Q 38:241–251
35. Kerkeni A, Behary N, Perwuelz A, Gupta D (2012) Dyeing of woven polyester fabric with curcumin: effect of dye concentrations and surface pre-activation using air atmospheric plasma and ultraviolet excimer treatment. Color Technol 128:223–229
36. Khan AA, Iqbal N, Adeel S, Azeem M, Batool F, Bhatti IA (2014) Extraction of natural dye from red calico leaves: gamma ray assisted improvements in colour strength and fastness properties. Dyes Pigm 103:50–54
37. Khan M, Khan M, Mohammad F (2004) Natural dyeing on wool with Tesu (flame of the forest), Dolu (Indian rhubarb) and Amaltas (*Cassia fistula*). Colourage 51:33–38
38. Khan MI, Ahmad A, Khan SA, Yusuf M, Shahid M, Manzoor N, Mohammad F (2011) Assessment of antimicrobial activity of catechu and its dyed substrate. J Clean Prod 19:1385–1394
39. Khan SA, Ahmad A, Khan MI, Yusuf M, Shahid M, Manzoor N, Mohammad F (2012) Antimicrobial activity of wool yarn dyed with *Rheum emodi* L. (Indian Rhubarb). Dyes Pigm 95:206–214
40. Khan SA, Khan MI, Yusuf M, Shahid M, Mohammad F, Khan MA (2011) Natural dye shades on woollen yarn dyed with Kamala (*Mallotus philippensis*) using eco-friendly metal mordants and their combination. Colourage 58:38
41. Kim TK, Yoon SH, Son YA (2004) Effect of reactive anionic agent on dyeing of cellulosic fibers with a berberine colorant. Dyes Pigm 60:121–127
42. Kirk RE, Othmer DF, Mark HF (1965) Encyclopedia of chemical technology. Interscience Publishers, New York
43. Knecht E (1941) A manual of dyeing for the use of practical dyers, manufacturers, students and all interested in the art of dyeing, vol 1, 9th edn. Griffin, London

44. Kongkachuichay P, Shitangkoon A, Chinwongamorn N (2002) Thermodynamics of adsorption of laccaic acid on silk. Dyes Pigm 53:179–185
45. Kumar JK, Sinha AK (2004) Resurgence of natural colourants: a holistic view. Nat Product Res 18:59–84
46. Lage M, Cantrell CL (2009) Quantification of saffron (*Crocus sativus* L.) metabolites crocins, picrocrocin and safranal for quality determination of the spice grown under different environmental Moroccan conditions. Sci Hortic 121:366–373
47. Loyson P (2010) Chemistry in the time of the pharaohs. J Chem Educ 88:146–150
48. Mayer F, Cook AH (1943) The chemistry of natural coloring matters: the constitutions, properties, and biological relations of the important natural pigments. Reinhold Publishing Corporation, New York
49. Mueller SO, Schmitt M, Dekant W, Stopper H, Schlatter J, Schreier P, Lutz WK (1999) Occurrence of emodin, chrysophanol and physcion in vegetables, herbs and liquors. Genotoxicity and anti-genotoxicity of the anthraquinones and of the whole plants. Food Chem Toxicol 37:481–491
50. Nourmohammadian F, Gholami MD (2008) An investigation of the dyeability of acrylic fiber via microwave irradiation. Prog Color Colorants Coat 1:57–63
51. Park JH, Gatewood BM, Ramaswamy GN (2005) Naturally occurring quinones and flavonoid dyes for wool: insect feeding deterrents. J Appl Polym Sci 98:322–328
52. Parvinzadeh M (2007) Effect of proteolytic enzyme on dyeing of wool with madder. Enzyme Microbial Technol 40:1719–1722
53. Perkin AG, Everest AE (1918) The natural organic colouring matters. Longmans, Green and Co., London
54. Rather LJ, Shahid I, Mohammad F (2015) Study on the application of *Acacia nilotica* natural dye to wool using fluorescence and FT-IR spectroscopy. Fibers Polym 16:1497–1505
55. Rattanaphani S, Chairat M, Bremner JB, Rattanaphani V (2007) An adsorption and thermodynamic study of lac dyeing on cotton pretreated with chitosan. Dyes Pigm 72:88–96
56. Rosenberg E (2008) Characterisation of historical organic dyestuffs by liquid chromatography-mass spectrometry. Anal Bioanal Chem 391:33–57
57. Samanta AK, Agarwal P (2009) Application of natural dyes on textiles. Indian J Fibre Text Res 34:384–399
58. Sarkar A (2004) An evaluation of UV protection imparted by cotton fabrics dyed with natural colorants. BMC Dermatol 4:15
59. Sewekov U (1988) Natural dyes- an alternative to synthetic dyes. Melliand Textilber 69: E145–E148
60. Shahid M, Islam S, Mohammad F (2013) Recent advancements in natural dye applications: a review. J Clean Prod 53:310–331
61. Shahmoradi Ghaheh F, Mortazavi SM, Alihosseini F, Fassihi A, Shams Nateri A, Abedi D (2014) Assessment of antibacterial activity of wool fabrics dyed with natural dyes. J Clean Prod 72:139–145
62. Singh R, Jain A, Panwar S, Gupta D, Khare SK (2005) Antimicrobial activity of some natural dyes. Dyes Pigm 66:99–102
63. Siva R (2007) Status of natural dyes and dye-yielding plants in India. Curr Sci 92:916–925
64. Son YA, Kim BS, Ravikumar K, Kim TK (2007) Berberine finishing for developing antimicrobial nylon 66 fibers: exhaustion, colorimetric analysis, antimicrobial study, and empirical modeling. J Appl Polym Sci 103:1175–1182
65. Van Stralen T (1993) Indigo, madder and marigold: a portfolio of colors from natural dyes. Interweave Press, Colorado, USA
66. Vera de Rosso V, Mercadante AZ (2009) Dyes in South America. In: Bechtold T, Mussak R (eds) Handbook of natural colorants. Wiley, Chichester, pp 53–64
67. Verhecken A (1989) Dyeing with kermes is still alive! J Soc Dyers Colour 105:389–391
68. Wang X, Li J, Fan Y, Jin X (2006) Present research on the composition and application of lac. Forest Stud China 8:65–69

69. Yusuf M, Ahmad A, Shahid M, Khan MI, Khan SA, Manzoor N, Mohammad F (2012) Assessment of colorimetric, antibacterial and antifungal properties of woollen yarn dyed with the extract of the leaves of henna (*Lawsonia inermis*). J Clean Prod 27:42–50
70. Yusuf M, Shahid M, Khan MI, Khan SA, Khan MA, Mohammad F (2015) Dyeing studies with henna and madder: a research on effect of tin (II) chloride mordant. J Saudi Chem Soc 19:64–72
71. Zarkogianni M, Mikropoulou E, Varella E, Tsatsaroni E (2011) Colour and fastness of natural dyes: revival of traditional dyeing techniques. Color Technol 127:18–27
72. Ziderman II (1990) "BA" guide to artifacts: seashells and ancient purple dyeing. Biblical Archaeol 33:98–101

Sustainable Dyeing and Finishing of Textiles Using Natural Ingredients and Water-Free Technologies

Kartick K. Samanta, S. Basak and S.K. Chattopadhyay

Abstract Wet-chemical processing of textile substrates starting from its preparatory to dyeing/printing followed by finishing is important for its value addition in terms of aesthetic value, removal of impurities, colour shade, colour pattern and requisite functionality. However, some of the traditional processes are water, energy and chemical intensive. In the recent time, due to global awareness on environmental pollution, climate change, global warming, carbon footprint and sustainability, both the academic research and industrial product development have been intensified to seek for sustainable dyeing and finishing processes, using biomacromolecules, biomaterials, plant extract, biopolymer and water-free technologies. In this context, deoxyribonucleic acid (DNA) from herring sperm, whey proteins, casein, chicken feather protein (CFP), banana pseudostem sap (BPS), spinach juice (SJ) and green coconut shell extract (GCSE) has been explored for improving the thermal stability of cellulosic, lignocellulosic and protein substrates. Similarly, agro-waste, nanolignin, silk sericin and aloe vera have been successfully extracted and applied in textile substrates to protect its user from the harmful ultraviolet (UV) rays. The above natural ingredients and a few more from marigold, manjistha, annatto, neem, turmeric, sandalwood, tulasi, jasmine, lemon, lavender and sandalwood have also been explored for natural dyeing, UV protective, aroma and antimicrobial finishing of textile. Water-less plasma and UV treatment can also be used as a pre-treatment, post-treatment, in situ reaction or post-polymerization for surface activation, oxidation, etching, polymerization, coating and deposition to impart value-added functionalities, such as water and oil absorbency, water and oil repellency, flame retardancy, UV protection, anti-static property and dyeing of various textile substrates.

K.K. Samanta (✉)
Chemical and Bio-chemical Processing Division, ICAR-National Institute
of Research on Jute and Allied Fibre Technology, Kolkata 700040, India
e-mail: karticksamanta@gmail.com

S. Basak · S.K. Chattopadhyay
ICAR-Central Institute for Research on Cotton Technology, Mumbai 400019, India

© Springer Science+Business Media Singapore 2017
S.S. Muthu (ed.), *Textiles and Clothing Sustainability*,
Textile Science and Clothing Technology,
DOI 10.1007/978-981-10-2185-5_4

Keywords Flame retardant · Biomacromolecules · Natural ingredients · Fragrance finish · Plasma and UV treatment · Sustainable · UV protective

1 Introduction

Chemical processing of a textile includes preparatory processing, coloration and functionalization, which are of significant importance so as to add requisite value to it. However, wet-chemical processing adopted by today's textile industry and small-scale process house consumes a large quantum of water, various chemicals and auxiliaries. Furthermore, they generate a large quantity of effluent that is finally discharged into the water stream contaminated with residual dyes, pigments and other chemicals. Some of the chemicals used in preparatory processing, dyeing and finishing of textile cause water, air and soil pollution, resulting in an adverse effect on the flora and fertility of agricultural land. In recent years due to increased awareness on human health and hygiene, while preserving natural resources, demand for natural fibre-based apparel and home textiles dyed and finished with natural ingredients, such as coloration with natural dyes, bio-polishing with enzymes, antimicrobial finishing with neem and aloe vera, ultraviolet (UV) protective by natural dyes and plant biomolecules, and flame retardant (FR) finishing by biomacromolecules, and plant extract is gaining both academic and commercial interest. Also, due to simultaneous concern on environmental pollution, climate change, global warming, carbon footprint and sustainability, efforts are also on to gradually replace part of the non-environment-friendly synthetic chemicals and auxiliaries with the eco-friendly plant- and vegetable-based natural ingredients, biomolecules, biomaterials and biopolymer. This will not only help to develop sustainable textile process and products for the traditional market, but also an array of new products for the upcoming niche markets, while preserving the natural resources.

The functional FR finishing of fibrous materials is an emerging area of research and sustainable product development that aims to protect not only its user, but also many valuable textile and non-textile products. In this context, the present chapter discusses in details the mechanism of FR finishing of cellulosic, lignocellulosic and protein substrates and recent development in FR formulations over the years, along with their merits and demerits. A special reference to the very recent developments on eco-friendly, cost-effective and sustainable FR formulations from plant ingredients, such as banana pseudostem sap (BPS), spinach juice (SJ), and green coconut shell extract (GCSE), and biomolecules, such as deoxyribonucleic acid (DNA), whey proteins, caseins, chicken feather protein (CFP) and hydrophobins along with their mechanism of action has also been reported. Besides flame retardancy, in the recent years, the consumers are also showing keen interest in the need of protection from the sun rays and UV rays that may induce skin damage due to the excessive exposure. The UV protective textile is important in our daily life, as about 6–7 % UV radiation is present in the solar spectrum reaching to the earth. Accordingly, the present chapter discusses in brief the UV radiation and its hazardous effects on the human body, followed by conventional methods of UV protection of natural as well as manmade textiles. It also describes the recent technological advancements in UV protection on cotton, silk, wool, linen, jute and other textiles using plant extracts, natural polymers and biomaterials in a cost-effective sustainable manner. Natural extracts of champa, lemon, sandalwood lavender, jasmine, rose, etc. containing active ingredients like santene, linalyl acetate, benzyl acetate, fusanol, teresantol, santalols, linalool and benzyl benzoate have been added in textiles by direct and microencapsulation technique for their mind-refreshing fragrance and skin nourishing attributes. Emerging water-free, sustainable textile processing using plasma and UV technologies as a pre-treatment, post-treatment, in situ reaction, and post-graft-polymerization for surface activation, oxidation, etching, polymerization, coating or deposition in order to impart various value-added functionalities, such as water and oil absorbency, water and oil repellency, FR, UV protection, anti-static, anti-felting, dyeing and printing are also discussed in details in the present chapter. As these high value textiles are produced mostly from the natural fibres and functionalized using natural ingredients (biomaterials/biomolecules) or water-free technologies, they can be effectively used for the production of sustainable textiles.

Sustainability is the current day's budding topic often used in the field of fibre, textile, fashion, agricultural, food and economics. The most common meaning of sustainability is the 'processes or products that can meet the today's needs of the society, without compromising resources for the future generations'. In fact, it means that something to sustainable it will be able to continue for a long time, without adversely effecting the environment, society or cost consistency. Therefore, in the context of sustainable textile and fashion, it is required to be produced without damaging the environment, people involved in the process and production, and without price escalation too high so that it will not economically feasible to produce anymore. Naturally coloured cotton, organic cotton, wild silk, organic wool, and even hemp and flax are the examples of important natural fibres in this category. For the production of textiles in a sustainable manner, the associated chemical processing as well as product development should also be sustainable in terms of cost, energy, water, applied chemicals, availability of raw material and biodegradability. As in the current practice high amount of water is used in wet-chemical processing of textile and some of the chemicals used in such process are not very much friendly to environment, human body and fabric's properties, therefore in the present chapter an emphasis has been given on dyeing and functional finishing of textile using water-less plasma and UV technologies, and using biomacromolecules and other plant-based natural ingredients.

The present chapter discusses the sustainable FR finishing of cellulosic, lignocellulosic and protein textiles and cellulosic paper substrate using agro-residues, such as banana sap, SJ and GCSE, and biomacromolecules, such as DNAs and proteins. These plant-based molecules and also mulberry fruit extract, natural dyes, chitosan, tulsi, silk sericin, nanolignin, aloe vera, honey, almonds, cucumber and mint have been applied in various textiles for UV protection, skin nourishing and infection control. In the context sustainability, the opportunity of emerging water-less plasma and UV irradiation for value-added finishing has also been summarized. The extracts of jasmine, lavender, champa, sandalwood and fragrances oils have been incorporated in the textile substrates to ensure a pleasant and fresh feel.

2 Sustainable Flame Retardant Finishing of Textile

2.1 Fibre and Flame

Among the various important functional finishing of textiles, FR finishing is important, as it is directly related to human health and material safety, and they are in market demand [1]. Cellulosic, lignocellulosic and protein textiles, such as cotton, flax, ramie, jute, silk and wool, are mostly used in apparels, home furnishing and other utility applications. Textiles meant for apparel and home furnishing, wholly cotton or cotton-blended textile are mostly preferred owing to its many positive attributes, like soft feel, good moisture regain, adequate thermal insulation, strength and service life. However, in many of such important applications, its usage is still restricted/limited, as it could catch flame easily. Indeed, cotton being a cellulosic fibre with lower limiting oxygen index (LOI) of 18 and low gross calorific value (GCV) of 16.4 MJ/kg catches flame readily and burns vigorously in an open atmosphere, making it difficult to extinguish that causes sometimes accidental death [2]. Though cotton and regenerated cellulose (viscose) have the combustion temperatures of 350 and 420 °C, respectively; however, in real time during burning, the temperature can reach to as high as 400–450 °C, leaving any residual mass. In the last five decades, both academic researchers and industrial product development have intensified in order to design and develop the suitable FR chemicals and application protocol in order to prevent the combustion of polymers/textiles or to reduce the flame propagation. This period can be considered as the 'Era of FRs' [3]. Their exponential growth in this field is arising mainly due to the large employment of polymers in applications, like transports, automotive, military, protective garments, furniture upholstery, hospital curtains and many more. As per the very recent report by Van der Veen and de Boer, the total consumption of FRs in Europe was 465,000 tonnes in 2006, 10 wt% of which were based on brominated compounds [3]. Cotton cellulose in contact with flame in nitrogen atmosphere pyrolyses through a single-step process, during which the cellulose degradation follows two competitive routes, namely depolymerization of the main chain to produce volatile species (levoglucosan, furan and furan derivatives) and dehydration with further auto-cross-linking to form a thermally stable carbonaceous structure, known as char [4]. Like natural fibres, thermoplastic synthetic fibres, such as polyester, nylon, acrylic and, even to some extent, polypropylene, are also used for the similar applications. Unlike the natural fibres, the situation is little bit complex in the thermoplastic fibres, as they most often shrink, melt and burn in contact with flame (heat). Though their combustion temperatures are considerably higher (450–550 °C), they being the thermoplastic in nature, first shrink, then soften and finally get melted in the temperature range of 165–265 °C. The shrinking of a fabric and dripping of molten liquid polymer can cause different degree of skin burning resulting in health injury. Natural as well as synthetic fibres, besides their apparel and home furnishing applications, are also used in hotels, hospitals, automobiles, railways and airways as tapestries and upholstery, where the FR finishing is also desirable. It may be noted that textile with an limiting oxygen LOI of ≤21 catches flame readily and burns profoundly in an open atmosphere. The samples with an LOI of ≥21 to ≤27 also catch flame and, however, burn slowly in an open atmosphere. On the other hand, the major requirement of a textile material to be considered as FR is that it should have an LOI value of more than 27. The burning situation is bit better in the case of lignocellulosic fibre, like jute with the LOI of around 21, making it ideal choice for packaging of agricultural crops and commodities, and upholstery and home furnishing applications. The situation is still better for the protein fibres, like wool and silk, as they possess higher combustion temperature of 600 °C (wool) along with higher LOI value of 23.5–25. As majority of the natural and man-made fibres/fabrics possess the LOI values of less than 25, they are required to be modified to improve their thermal stability either during the manufacturing process (e.g. polymerization, fibre spinning) or in their final stage of fabric finishing, making them acceptable for the requisite end applications. Fire retardant finishing of a textile substrate deals with the combustion process of that particular fibre in contact with thermal energy in the form of flame. It makes the differences between the samples before and after the FR treatment in terms of LOI, burning rate, heat of combustion, char length, flame time, afterglow time, rate of heat release, production of char and tar, and formation of flammable and non-flammable gases. Therefore, during the formulation of an efficient FR, certain points are to be considered, such as how to enhance the pyrolysis temperature of the fibre, decrease the formation of flammable gases, mask the availability of oxygen and enhance formation of char and non-flammable gases.

2.2 Traditional Flame Retardant Finishing of Textile

The FR finishing of textiles can be considered as durable, semi-durable or non-durable based on its performance efficacy after washing. In the last few decades, both academic and industrial researchers have confined to design suitable chemicals to prevent or delay the combustion process of fibrous materials, especially textile fibres/fabrics [5]. Cellulose pyrolyses in nitrogen environment in two alternative pathways, which involve the decomposition of the glycosyl units to char at lower temperature and the depolymerization of such units to volatile species at higher temperature [6]. Some of the well-explored non-durable to durable FR chemicals available in the market are inorganic salts, borax and boric acid mixture, diammonium phosphate, urea, halogens and halogen-based derivatives, phosphorus, nitrogen, aluminium, magnesium, boron, antimony, molybdenum and the recently developed nanoparticles, nanofillers and nanoclay [3, 7–9]. In the sixteenth to eighteenth centuries, the use of clay, gypsum pigments, borax, alum, ferrous sulphate, ammonium phosphate and ammonium chloride has been reported for FR finishing of jute and cotton textiles, and other non-textile substrates [10, 11]. Inorganic salts are also known for imparting flame retardancy to the cellulosic textiles for a long time that will not come direct in contact with liquid in the forms of water, rain or perspiration [11]. As far as cellulosic substrates are concerned, phosphorus- and nitrogen-based FRs have been proven to be the best performing systems currently used.

The bromine and chlorinated halogen compounds either alone or in combination with admixture of antimony could ensure synergism, and the same were introduced in the market around middle of the nineteenth century. Halogenated FRs though have very good fire retardant efficacy and have some environment limitations [1]. Nowadays, the production and usage of halogenated FRs derivatives are restricted more and more by the European Union and they have been voluntarily banned in the USA [3]. Some of the widely used molecules employed in the formulation of halogen-based FRs, such as pentabromodiphenyl ether, decabromodiphenyl ether (or oxide) and polychlorinated biphenyls, have been found to be persistent, bio-accumulative and/or environmentally toxic to animals and humans.

FRs based on phosphorous, nitrogen and halogen, like N-alkyl phosphopropi-onamide derivatives and tetrakis phosphonium salt, have been widely explored for commercial end use [8, 12]. Many of these FRs were found to reduce the fabric's tensile and tear strength noticeably, besides making the fabric stiffer, mainly due to the requirement of acidic pH during application as well as involvement of high temperature drying/curing process. In addition, the process was found to be toxic and costly [13]. Due to the above adverse effects of halogenated compounds on environment, the halogen-free compounds are being developed rapidly to fulfil the increasing market demand of environment-friendly formulation. In this line, phosphorus-based commercial available FRs compounds may be consider as a suitable alternative to halogen-based derivatives [5]. Additionally, a new mecha-nism of flame retardancy, i.e. intumescent technology, has also introduced in the market in 1996. From an overall point of view, intumescent coatings represent a suitable way for conferring FR to different substrates, like metals, plastics, foams, polymer and textiles. It efficacy originates from the presence in their chemical structure of a carbon source, an acid donor and a blowing agent [14]. In this direction, carbonizing compounds, such as polyphenols, carbohydrates, polyhydric alcohols and resins, dehydrating compounds like urea, diammonium phosphate, ammonium sulphate and melamine, and foam-forming agents like dicyandiamide and melamine have been utilized to increase more carbonaceous char formation [15]. The intumescence characteristics of a material also depend on the atomic ratio of carbon, nitrogen and phosphorous [8]. Some of the important intumescent pro-cesses reported in literature are as follows: (i) ammonium polyphosphate (APP) for coating of rigid polyurethane foams, (ii) melamine (nitrogen containing) for var-nishes and paints, (iii) ammonium polyphosphate and (iv) ammonium pentaborate.

In the same principle, sodium metasilicate nonahydrate was utilized to make the jute textile FR and antimicrobial [16, 17].

The most efficient FR systems presently available for cellulosic cotton textile are phosphorus-containing species that release phosphorus acids at elevated temperature, which act as Lewis acids and promote the char formation, while favouring the dehydration of cellulose and inhibiting its depolymerization [18]. Phosphorus–nitrogen inflating FR emits less smoke, and the poisonous and corrosive gas, preventing molten dropping, making it an ideal textile application [1]. Diammonium phosphate along with ammonium salts and urea acts in condensed phase mechanism and was also used as a fire retardant for cellulosic textiles. Such compound can reduce the pyrolysis temperature of cellulose, while promoting the production of more char and less flammable volatiles. In this context, the synergistic upshot of trimethyl melamine (TMM) and dimethyl dihydroxy ethylene urea (DMDHEU) as a nitrogen provider with organophosphorous compound was also explored [19]. In the last few decades, halogen-free FRs based on the composition nitrogen, phosphorous and halogen like tetrakis phosphonium salt and N-alkyl phosphopropionamide derivatives have also been widely used for coatings and back-coated textiles in commercial scale [5, 15]. Indeed, the benchmarks in terms of having acceptable costs and meeting the current health, safety and environmental issues are showing acceptable technical performances and satisfying the flammability regulatory legislation of Proban® and Pyrovatex® processes [18]. Presently, most of the textile industries use N-methylol dimethylphophopropionamide (Pyrovatex) with melamine resin for fire retardant finishing of cotton textile.

2.3 Sustainable Flame Retardant Finishing Using Natural Ingredients

As reported earlier, many of the present-day FR formulations, mainly those are based on phosphorus and nitrogen, antimony-halogen, diammonium phosphate, THPC, Pyrovatex, urea and melamine–formaldehyde, have adverse effect on the environment, human health and fabric's physical, mechanical and aesthetic properties. Some of the environment-friendly chemicals, like sodium metasilicate nonahydrate, BTCA, BTCA-hydroxyalkyl organophosphorous compound, organophosphoramidates, Noflan and a few more, have also been explored in this direction. Water-less emerging plasma technology, layer-by-layer deposition, nanomaterials and sol–gel processes were also explored in the recent time as promising fire retardant solutions with an aim to reduce the number of processing steps, usage of water, quantum of chemicals required and cost of production. In spite of such developments, there is a need of apposite FR formulation, preferably those derived from the renewable plant and biological sources to make the process simple, environment-friendly, cost-effective, suitable for both natural and synthetic fibres and limited to no adverse effect on the fabric physical and mechanical properties with requisite durability for sustainable FR finishing of fibrous materials. In this direction, the usage of biomacromolecules or biomaterials, like proteins, nucleic or DNAs, whey protein isolate (WPI), CFP, caseins and hydrophobins, may represent a worthy alternative to the traditional approaches in the context of green, novel, natural, sustainable and cost-effective FR solution for cellulosic cotton textile as reported below. Similarly, plant molecule (i.e. plant extract)-based formulation like BPS, SJ and GCSE have also been emerged in the recent time for dyeing, and functional FR and UV protective finishing of cellulosic, lignocellulosic and protein substrates as they possess active ingredients like phosphorous, chlorine, silicate and many other metallic compounds [2, 20–26]. Moreover, these agro-wastes are abundantly available in many countries and could be considered as green, sustainable and cost-effective FR for textiles and paper substrates as summarized below.

DNA is the carrier of genetic information and gene expression. In the recent time, Bosco et al. have demonstrated that this biomacromolecule from herring sperm and salmon fish is very effective in providing FR functionality in cellulosic cotton fabrics [4, 18, 27, 28]. The complex double helix of DNA represents a potential and intrinsic intumescent-like FR attributes, as it is composed of three typical components of an intumescent formulation, namely the phosphate groups that provide phosphoric acid, the deoxyribose units that act as a carbon source, and blowing agents (upon heating a (poly)saccharide dehydrates forming char and releasing water) and the nitrogen-containing bases (guanine, adenine, thymine and cytosine) that may release ammonia [18]. These molecules altogether are responsible in the formation of more carbonaceous char and release of ammonia to enhance the thermal stability of cotton fabric. This process occurs only if the DNA could absorb heat from the surrounding zone, and hence it should be responsible for a strong temperature reduction during the combustion process. The effect of different DNA add-ons, namely 5, 10 and 19 wt%, has been thoroughly investigated as far as the flammability and the resistance to an irradiating heat flux of 35 or 50 kW/m^2 are concerned. It was observed that 10 wt% add-on was the minimum requirement to make the cotton textile self-extinguishable, when a methane flame was applied for the test [18]. The process postulates that phosphate groups of DNA behave similarly to an inorganic phosphate salt, like APP, a common FR chemical for cellulosic substrates. After application of DNA molecules, the LOI values were found to increase from 18 in the untreated sample to 23 (5 % add-on), 25 (10 % add-on) and 28 % (19 % add-on) treated samples as reported in Table 1. This finding clearly depicts that the DNA coating can be considered as an efficient FR formulation. The same research group has also investigated the effects of DNA molecular size, pH of the DNA aqueous solutions and numbers of impregnations on the thermal stability of cotton fabric. Herring DNA, at three different molecular sizes namely, low (100–200 base pairs, bp), medium (400–800 bp) and high (2000–10,000 bp), was selected as model biomacromolecule, and after application, the improved thermal stability of the cotton fabric was analysed through horizontal flame spread test and cone calorimetry analysis with an irradiative heat flux of 35 kW/m^2 [4]. It was interesting to note that in spite of similar chemical composition, the coating containing low molecular size DNA as well as multiple impregnations leads to achievement of superior fire

Table 1 Flammability parameters in untreated and natural ingredient-treated cellulosic, ligno-cellulosic and protein substrates [1–5, 18, 20–28]

Sl. no.	Type of samples	LOI	Add-on (%)	Burning rate (mm/min)	Total burning time (s)[b]	Char length (cm)
1	Untreated cotton fabric	18	–	75	60 (60 + 0)	–
	BPS-treated cotton fabric	30	4.5	7.5	904 (4 + 900)	–
2	Untreated jute fabric	21	–	62	60 (60 + 0)	–
	BPS-treated jute fabric	33	3	15	605 (5 + 600)	–
3	Untreated paper	18	–	280	50 (40 + 10)	–
	BPS-treated paper	28	6	30	500 (0 + 500)	No
	Bio-enriched BPS-treated paper	33	8	–	45 (0 + 45[a])	5.3
4	Untreated paper	18	–	187	90 (60 + 30)	–
	GCSE-treated paper	27	5	82	210 (60 + 150)	–
	GCSE + Boric-treated paper	33	9	–	0 (0[a] + 0[a])	2
5	Untreated wool fabric	25	–	–	90 (90 + 0)	–
	BPS-treated wool fabric	35	5	–	3 (0[a] + 3[a])	6
6	Untreated cotton fabric	18	–	250	60 (60 + 0)	–
	Spinach juice (SJ)-treated cotton fabric	30	8	37.5	404 (4 + 400)	–
7	Untreated cotton fabric	18	–	90 (total burning rate)	66 (66 + 0)	–
	DNA-treated cotton	28	19	180 (total burning rate)	2	0.6
8	Untreated cotton fabric	–	–	90	78	–
	Whey protein-treated cotton	–	25	60	126	–
9	Untreated cotton fabric		–	90	72	
	Casein-treated cotton		20	60	100	
	Hydrophobin-treated cotton		20	66	104	
10	Untreated cotton fabric	18	–	–	22 (12 + 10)	>30
	Chicken feather protein-treated cotton	30.1	5.72	–	187 (7 + 180)	>30
	Chicken feather protein + borax + boric acid-treated cotton	39.9	8.10	–	2 (0 + 2)	4.5

[a]Sample does not catch flame or afterglow is stopped
[b]Total burning time = burning with flame time + burning with afterglow time

protection on textile substrate. Indeed, multiple impregnation at pH 4 and 8 yielded self-extinction of 86 and 74 % tested samples, respectively. Under an irradiative heat flux of 35 kW/m^2 with multiple impregnations, despite pH of the applied solutions, all the treated samples could exhibit a profound decrease in total heat release and heat release rate peak as far as flame retardancy is concerned. In the same direction, the DNA was also combined with chitosan and applied in cotton fabric layer-by-layer assemblies [14]. Cotton textile with 5, 10 and 20 DNA/chitosan bi-layers resulted in dry weight gain of 2, 5, 7 and 14 wt%, respectively. Among these various samples, the 20 bi-layer coatings were found to be capable of reaching the self-extinction of cotton textile in horizontal flammability tests with increasing LOI value up to 24 %, while decreasing the heat release rate by 40 % in cone calorimetry test.

Similar to FR finishing of cotton textile using DNA biomacromolecule, efforts were also directed to make the cotton textile FR using folded and unfolded whey proteins, casein and hydrophobins, as these are rich in phosphate, disulphide and protein that can influence the pyrolysis of cellulose polymer by an early char formation [18]. WPI powder (93.5 wt% protein) is composed of lipids (ca. 0.5 wt%), carbohydrates (ca. 1 wt%), ash (ca. 2.2 wt%) and moisture (ca. 2.8 wt%); it also has amino acid content equal to 89.5 wt% [27]. Cellulosic cotton fabrics were impregnated in the folded and unfolded protein suspensions for 1 min at 30 °C to form a continuous and coherent film on the fabric structure, although some cracks were visible, where proteins remained still folded. The elemental analysis of the COT_WP (cotton with folded protein)- and COT_ DWP (cotton with unfolded protein)-coated samples depicts the presence of carbon (C), oxygen (O), nitrogen (N) and sulphur (S) as main constituents. The applied coating was responsible to partially protect the cotton textile from flame probably by hindering the oxygen diffusion and absorbing heat during the combustion process. Alongi et al. [5] have also recently reported a green FR finishing of cotton fabric using caseins and hydrophobins. As these proteins are rich in phosphorus and sulphur, they were utilized for homogeneous deposition on cotton fabrics. The active ingredients in such proteins are able to favour the dehydration of cellulose instead of its depolymerization that strongly enhanced the production of thermally stable carbonaceous char; thus, flame retardancy was achieved. The untreated sample in contact with a methane flame for 3 s burns vigorous, while producing no residual mass. In contrary, the caseins or hydrophobin-coated fabrics showed profound increase in the total burning time (+40 %), while simultaneous decrease in the total burning rate (−35 %). Like above, Wang et al. [1] have reported incorporation of CFP for making cotton textile FR. Chicken feather, a biodegradable waste protein material abundantly available in a large quantity throughout the world, can be used as an eco-friendly FR formulation for textile substrates. CFP-based phosphorus–nitrogen-containing FR was synthesized with CFP, melamine, sodium pyrophosphate and glyoxal. The flame retardancy efficacy of such formulation was evaluated in combination with borax and boric acid. Results implied that all these compounds could exhibit good synergisms and facilitate to form a homogenous and compact intumescing char layer, and thus flame retardancy was enhanced.

Like above, i.e. FR finishing of textile using DNA, whey proteins, CFP, casein and hydrophobins, the BPS, an agro-waste product available in large quantity in many countries, was applied to a pre-mordanted (5 % tannic acid +10 % alum) cellulosic and lignocellulosic textiles by pad-dry method. The untreated cotton and jute fabrics have the LOI values of 18 and 21, respectively, as shown in Table 1. When both the samples were treated with the alkaline BPS solution, the LOI value was found to increase more than 27, and hence, the textiles can be considered as FR textiles [20, 21]. The BPS-treated cotton fabric burns with flame only for 4 s; thereafter flame is stopped and the sample continues to burn with an afterglow for 900 s. On the other hand, untreated sample burnt completely within 60 s with high flame intensity. It was observed that the add-on percentage and the LOI values remain alike, when BPS was applied in acidic and neutral pH, whereas it could exhibit better thermal stability in alkaline condition [22]. As expected, the rate of heat production was more in the case of untreated sample, then the BPS-treated sample. It may be noted that in the treated sample, there was much more time (904 s) available, either to escape from the fire place or to extinguish the fire as compared to only 60 s available for the untreated sample. Unlike the cotton fabric, the jute sample showed better thermal stability in the presence of BPS. It was also interesting to note that when the BPS liquid biomass was concentrated by 100 and 400 %, and the same was applied, the LOI value increased to 36 and 40, respectively [21]. Indeed, these samples did not catch flame at all. In the treated sample, the observed burning rate was almost 15 times lower as compared to untreated sample. The treated fabric possesses a LOI value of as high as 33 and flame time only 5 s, compared to the untreated sample, where the respective values were 21 and 60 s. Though the flame stopped within 5 s in the treated sample, it could burn slowly at a rate of 24.8 mm/min. The advantages of the plant-based FR formulation are as follows: (i) flame was self-extinguishable in short period of time and (ii) the application process is quite simple and straightforward. There was no significant adverse effect (<5 % change) either on fabric's tensile or tear strength in the FR-treated samples, whereas most of the FR finish can cause 10–30 % reduction in tensile strength [13]. The flammable characteristics in the untreated and the BPS-, SJ-, GCSE-, DNA-, whey protein-, CFP-, casein- and hydrophobin-treated cellulosic, lignocellulosic and protein textiles, and paper are shown in Fig. 1 and Table 1.

In a FR textile, the combustion with afterglow is not as serious as combustion with flame, where the maximum temperature reaches up to 350 °C. The temperature generated during the burning of untreated cotton and jute fabrics was as high as 395–400 and 380 °C, respectively. On the other hand, the maximum temperature measured during the burning of the BPS-treated samples with afterglow was found to gradually decrease with the burning time, initially 320 °C, followed by 280, 250 and finally 120 °C at an interval of 60 s. The burning rate measured in horizontal flammability tester was found to decrease from 75 to 7.5 mm/min, which is 10 times lower than that of control sample. The scanning electron microscope

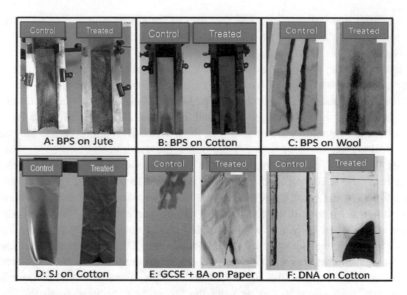

Fig. 1 Burning behaviour of the control, and the BPS-, SJ-, GCSE- and DNA-treated cellulosic, ligno-cellulosic and protein substrates [2, 4, 21, 22, 24, 25]

(SEM) of the BPS-treated sample represents an intact closed-cell char morphology containing many small pockets of gases. Possibly, BPS works as an additive intumescent FR that upon heating swells and forms a protective thick honeycomb like coating on the textile substrates, thus inhibiting the flow of volatile into the flame zone. The coating also protects the remaining part of the polymer from the thermal decomposition, while ensuring more char formation. On the other hand, poor char morphology with prominent presence of channels was visible in the control sample that acts as a pathway of volatile transport [20]. However, the efficacy of the finish gets reduced a little bit, i.e. LOI from 30 to 24 after single washing as per ISO 1 method. The gross calorific value (GCV) measured in an oxygen bomb calorimeter showed a value of 16,299 J/g for the control jute fabric and the same value was only 12,848 J/g in the BPS-treated sample (concentrated), resulting in almost 21 % reduction in heat production [21]. The BPS-treated cotton also follows a similar trend, where GCV value reduced from 16.4 to 13.1 MJ/kg in the untreated to treated samples, respectively [2]. Similar to the application of BPS in the cotton fabric, the LOI value also reduced from 40 to 30 in the jute fabric after single standard washing.

The cone calorimeter analysis of the untreated and the BPS-treated cotton fabrics was performed to elucidate the interaction of flame with fibre in terms of mass loss rate, heat release rate and amount of heat generated as summarized in Table 2. The length of time observed for ignition in the control and the treated samples are 7 and 14 s, respectively. It depicted that the BPS treatment on cotton fabric increased the sample ignition time as well as delayed the time to flame out. The mean heat release rate and the total heat release in the untreated cotton fabric were 31 and 29 % higher

Table 2 Summary of cone calorimeter results of the untreated and BPS-treated cotton fabrics at a heat flux of 35 KW/m^2 [22]

Heat release-related results					
Samples	H.R. R^a (KW/m^2)	T.H.R^b (MJ/m^2)	E.H.C^c Peak (MJ/Kg)	E.H.C (MJ/Kg)	MARHE^d
Untreated	34.0	5.24	40.4	21.5	133.2
BPS treated	23.5	3.72	27.1	12.3	109.5

Mass loss and smoke-related results					
Samples	Average specific mass loss rate (10–90 %) g/m^{2}s	Initial mass (g)	Sample mass loss (g/m^2)	Total smoke release (m^2/m^2)	CO$_2$ release (Kg/Kg)
Untreated	10.2	2.00	229.8	4.83	1.08
BPS treated	5.59	3.16	296.2	7.33	0.76

Ignition-related results			
Samples	Time to ignition (s)	Time to flameout (s)	Mass at flameout (g)
Untreated	7.00	20.0	4.72
BPS treated	14.0	33.0	5.95

Note ^{a}*H.R.R* Heat release rate
^{b}*T.H.R* Total heat release
^{c}*E.H.C* Effective heat of combustion
^{d}*MARHE* Maximum average rate of heat emission

as compared to BPS-treated cotton, respectively [22]. In the heat of combustion study, the control fabric showed an average combustion heat of 21.45 MJ/Kg, whereas the BPS-treated fabric exhibited only 12.24 MJ/Kg, which is almost half than that of the control sample. Indeed, the heat of combustion peak observed for the control fabric was at 40.44 MJ/Kg, whereas the same was 27.12 MJ/Kg for the treated fabric. The maximum average rate of heat emission (MARHE) was 17.8 % less in the treated sample as compared to the untreated sample. Based on the LOI value, afterglow, TGA analysis, smouldering and smoke formation, the burning models of the control and treated cotton fabrics have been proposed (Fig. 2). Treated fabric showed a slow thermal decomposition without any flame and the temperature generated during the thermal degradation was also less (200 °C) than the observed values in the control sample.

Similar to the application of BPS in cotton and jute textiles to improve the thermal stability, the efficacy of bio-enriched-BPS was also evaluated on cellulosic hand-made paper of 200 GSM (areal density). The sample was pre-mordanted with tannic acid and alum, like as reported above for textiles, prior to application of BPS. Both the control and the mordanted papers showed an oxygen index of 18. The oxygen index values were found to increase to 28 and 33 in the normal BPS and the bio-enriched BPS-treated papers, respectively [23]. As a consequence of this, the burning rate gets reduced markedly from 280 to 30 mm/min in the untreated to treated samples, respectively. The better thermal stability in the bio-enriched

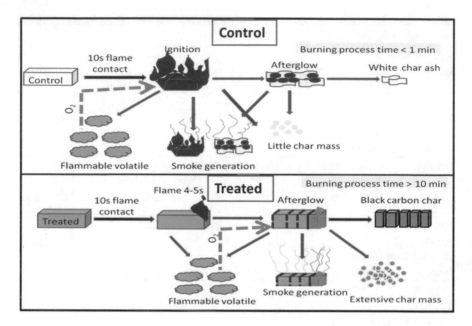

Fig. 2 Vertical burning model of control and BPS-treated cotton fabrics [22]

BPS-treated paper over normal BPS-treated paper was possibly due to the presence of more quantity of active FR ingredients, like Si, Cl, K and Ca as detected in the energy dispersive X-ray (EDX) analysis.

Protein fibre (e.g. wool) has a better heat retention capacity and thermal resistance property, and hence it can be preferably used as thermal resistance textile for the personnel working in oil handling, fire fighting, as a furnace attendance and in making winter dress materials [24]. As wool fibre contains high amounts of nitrogen, water and sulphur, it is to some extent inherently FR and has an ignition temperature of 500–600 °C. Its higher LOI value of 25 makes it a preferably better choice over the other cellulosic and lignocellulosic natural fibres, as far as the flame retardancy is concerned. However, fire retardant finishing of wool fibre is also recommended in cases, where certain prescribed flammability standards for specific end use are required to be satisfied. Similar to FR finishing of cotton, jute and paper with BPS, the woollen textile was also dyed and finished with a BPS liquid in acidic condition. In vertical flammability tests, the control wool fabric was found to burn completely with a flame within 80 s, whereas the treated woollen fabric at a neutral pH of BPS solution showed no flame and no afterglow, except a char length of 6 cm [24]. When BPS was applied in acidic pH of 4.5 and 5.5, the treated wool fabrics were found to burn with a black colour, forming a heavy chars mass. During heating, the control wool fabric showed a rapid shrinkage and formed a light weight fragile ashy black colour char mass, which, during breaking into smaller pieces, released lots of water. On the other hand, the BPS-treated wool fabric at a neutral pH of 7 showed less shrinkage and yielded a harder and more blackish carbonaceous char mass. The treated sample could also maintain a lower degree of damage of fibres, with some visibility of scale morphology in SEM analysis. After application, there was only 6 % reduction in tensile strength compared to untreated woollen fabric. After single washing as per ISO 1 method, the treated fabric could secure 85.8 % retention of the initial LOI value. The woollen fabric was dyed with 2 % Acid Telon Blue 25 dye using the BPS liquid as a dyeing medium and the results were compared with those samples dyed in water medium [24]. The fabric exhibited more colour exhaustion, colour strength and thermal stability as compared to the control wool, when it was dyed in acidic condition (pII 5.5) of the BPS solution. The BPS treatment showed no significant improvement in colour strength of the wool textile up to pH 5.5. All the treated fabrics (with varying pH) exhibited a reddish tone after the treatment. Unlike the dyeing of wool in a water medium, when the fabric was dyed in BPS bath, it showed no significant change in colour strength value at both the pH of 4.5 and 5.5. In flammability test, the wool fabric dyed from the BPS bath (pH 5.5) showed an improved oxygen index of 34 as compared to LOI value of 26 in the water medium-based dyed fabric at same pH.

Sustainable FR functionality was also imparted to cellulosic cotton textile using another vegetable-based extract, i.e. spinach leaves juice (SJ) [26]. The juice after extraction was converted to alkaline and then applied in the cellulosic cotton fabrics. The treated fabrics showed good FR property with LOI value of about 30 compared to the control fabric that has an LOI of only 18. As a result of this, the treated fabric did not catch flame. In horizontal burning test, the treated sample was found to burn with afterglow (whcn flame is stopped) at a flame propagation rate of 10 mm/min, which was almost several times slower than the untreated sample. The thermal degradation and pyrolysis of both the control and the SJ-finished fabrics were analysed by thermogravimetric analysis. The treated fabric could also produce a good natural green colour, which is stable even after a washing cycle. As expected, there was insignificant change in other physical properties of the fabric on application of SJ. The improvement in thermal stability was due to the existence of mineral salts in the form of chlorides, such as magnesium chloride, sodium chloride and sodium silicate [26].

In continuation of the application of BPS and SJ on cellulosic and lignocellulosic textiles, and on paper substrate for improvement of their thermal stability, the present authors have also reported the application of another natural liquid biomass, i.e. green coconut shell extract (GCSE) in combination with boric acid (BA) for improving the thermal stability of cellulosic paper. The GCSE, a waste product available abundantly in many countries, was used as a sustainable novel and natural FR agent [25]. The GCSE-treated papers with or without the presence of 2 % BA showed the LOI values of 27 and 33, respectively, whereas untreated paper showed an LOI of 18. The burning rate gets reduced from 187 mm/min to almost non-flammable attributes in the untreated to treated samples (Fig. 1). The improved thermal stability was possible owing to the presence of phosphorous-based salts, silicate, chlorinated compounds and other positive metallic salts based on potassium, zinc, copper and magnesium that probably had catalysed in dehydration and char formation. GCSE possibly acts in a condensed phase, providing a blanket effect surrounding the treated paper, while replacing the available oxygen [25]. Similar to the dyeing of protein wool fibre in the BPS medium, cellulosic paper was also dyed with acid and basic dyes in GCSE liquid medium and the results were compared with those samples dyed in a water medium. It was observed that the acid dyed cellulosic substrates in the GCSE medium were darker than those dyed with water medium. However, an opposite phenomenon was noticed in the case of basic dyed samples. This implies that GCSE contains mostly the positive metal ions that act as a mordant on the paper surface and attract the negatively charged acid dye molecules.

2.4   Chemical Composition of Natural Ingredients and Their Mechanism of Action

The FR characteristics by application of BPS may be due to the presence of phosphate and other mineral salts. Positive and negative secondary ion mass spectra (SIMS) were used to identify the presence of different metal chloride and phosphate. In addition, energy dispersive X-ray spectra analysis depicts the existence of magnesium, sodium, aluminium, phosphorous, potassium, chlorine and calcium. Fourier transform infrared spectroscopy analysis depicted the presence of inorganic salts in the pure BPS sample [2]. A recent analysis of liquid BPS also showed the presence of sodium chloride, potassium chloride and metal phosphate [29]. X-ray fluorescence studies (XRF) of extracted liquid BPS showed the presence of elements like phosphorous, chlorine, silicon and trace amount of aluminium. The alkaline BPS-treated fabric showed elements like chlorine, phosphorous, silicon, calcium and aluminium, where chlorine and phosphorous were predominant in nature [22]. In contrary, the improved thermal stability in the SJ-treated fabric appeared to be the presence of phosphate and silicate, and the same has also been identified by elemental and spectroscopic analysis. From the various instrumental results, it was postulated that the chloride, silicate, phosphate and other mineral salts present in BPS and SJ possibly have increased the thermal resistance of the cellulosic and lignocellulosic textiles by forming less flammable gases and more char. It has also been reported in literature that the presence of organometallic additives in cellulosic textiles could increase the char formation, while reducing formation of tar [30]. From nitrogen and phosphorous analysis, it was observed that with increasing pH, keeping the BPS concentration constant, there was a gradual increase in add-on percentage, resulting in a proportional increase in nitrogen and phosphorous content on the fabric surfaces. The nitrogen and phosphorous contents were estimated to be 1.7 % for the pH 10 solution-treated sample, and the corresponding LOI was 30. The presence of various inorganic salts was roughly observed in the FTIR analysis and estimated by the EDX elemental analysis of BPS [20]. Secondary ion mass spectroscopy (SIMS) analysis has confirmed and relatively quantified the amount of phosphate, phosphite, chloride, potassium chloride and other metal elements exist in the pure BPS [22]. The negative SIMS of BPS showed the presence of major molecules at different mass units, such as H^- (1 amu), C^- (12 amu), CH^- (13 amu), N^- (14 amu), O^- (16 amu), OH^- (17 amu), F^- (19 amu), Cl^- (35, 37 amu), PO_2^- (62, 63 amu), PO_3^- (79 amu), KCl^- (74, 76 amu), Cl_2^- (70 amu) and many more. The positive SIMS of the same sample showed the presence of various metal ions, like Mg^+ (24, 25 amu), K^+ (40, 41 amu), Mn^+ (55), Fe^+ (55, 56) and a few more. Due to the presence of such active ingredients in the BPS-treated jute and cotton textiles, samples could show better thermal stability [20].

In the BPS-based FR finishing, its coating may act as an intumescent that swells on heating and increases its volume, thus protecting the underlying cotton textile from the heat. SEM picture of the treated fabric revealed the presence of thick protective coating on the fibre surface. In contact with flame, it favoured the formation of a stable protective char that could hinder the exchange of oxygen and other combustible volatile. On the other hand, SEM image of the char of control paper showed a net like open capillary structure, through which the flammable volatile could easily escape during burning, while ensuring continuous burning of a paper. On the other hand, SEM images of chars of the BPS and the bio-enriched BPS-treated papers showed an intact honeycombed-like closed-cell structure that contains pockets of gases [23]. This closed-cell char morphology restricts the flow of flammable volatiles during burning, as shown schematically in Fig. 3. During

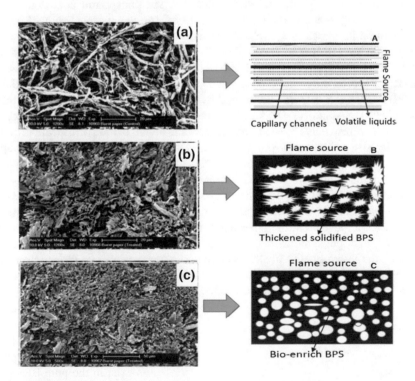

Fig. 3 SEM pictures of chars and models of control **a**, BPS-treated **b** and bio-enriched BPS-treated **c** papers during burning [23]

pyrolysis or decomposition of cellulose produces hydrocarbon fragments that along with flammable gases could migrate towards the flame source, where cyclization and aromatization lead to the formation of carbonaceous soot. Pyrolysis/decomposition gets started earlier in BPS-treated paper, resulting in an extensive dehydration and formation of additional hydrocarbon mass that might have partially migrated through the closed-cell char structure, and the same may be regarded as the presence of afterglow [23]. In the bio-enriched BPS-treated paper, the formation of more closed-cell char morphology might have hindered the transport of flammable volatiles and hydrocarbons, restricting burning not to happen, except the presence of a little bit glow, which too was also self-extinguishable in a short while.

In the FR finishing of cotton textile with DNA biomacromolecule, the DNA is considered as a polymer, which consists of two long chains of nitrogen-containing bases, adenine (A), guanine (G), cytosine (C) and thymine (T), allocated in the inner portion of the skeleton, with backbones made of five-carbon sugars (deoxyribose units) and phosphate groups, placed outside. This biomacromolecule acts as an intumescent-like mechanism. As reported earlier that the efficacy of the commonly used intumescent FR systems depends due to the presence of a carbon source, an acid donor and a blowing agent in their structure [14]. The DNA macromolecules naturally contain all these three active ingredients. The phosphate groups, involved in phosphodiester bonds, are the phosphoric acid donors, the deoxyribose sugars are carbon sources, and blowing agents together with purine and pyrimidine bases may release ammonia within 180–230 °C, during the pre-ignition step. In order to elucidate the role of molecular size of the different DNAs, the phosphorus content of HS-DNA and HT-DNA, as well as the fabrics coated with three different solutions of DNAs (HS-DNA 2.5 %, HT-DNA 0.5 % and FHT-DNA 0.5 %), was performed by ICP-MS analyses. Result demonstrated that the HS-DNA and HT-DNA possess the almost similar phosphorous content of 8 % and the respective coated fabrics contain about 0.5 %, irrespective of the molecular size of the employed DNA. The DNA is believed to decompose upon heating, producing an intumescent characteristic that is responsible for the formation of a multicellular, foamed and thermally insulating material at a relatively low temperature (120–250 °C), ultimately ensuring FR pathway. By examining the char morphology after the samples were burnt, the possible mechanism of FR characteristic was proposed. On the other hand, the caseins are able to generate globular structures like phosphorus-rich bubbles that blow up during the combustion, whereas the hydrophobin-treated samples showed several un-blown pearl-like bubbles that still remain intact. This performance was attributed to the cleavage of the disulphide bonds and further on to the cross-linking of amide groups [5].

3 Sustainable Textile Processing Using Water-Free Technologies

3.1 Application of Plasma Technology in Textile

Different types of irradiation techniques, like plasma, laser, gamma and UV rays, are being used nowadays, as an alternative to the water-based chemical processing of textiles [31]. Textile chemical processing is desirable for its improvement in functional and aesthetic properties, while making it suitable for dyeing and printing. The traditional wet-chemical processes of textile are water, chemical and energy intensive owing to multistep operations, like padding, drying, curing and post-washing. In the last five decades, several technological advancements have been realized in the textile processing arena with the aim to reduce cost of production or environmental impact, such as digital printing, spray and foam finishing, application of enzymes, natural dyes, use of biomaterials, plant extracts and biomacromolecules, ultrasound assisted dyeing and dispersion, low material to liquid processing, infrared dyeing and drying, and water-free plasma processing [32–34]. Plasma is a partially ionized gas composed of positive and negative ions, electrons, excited molecules, photons, UV light and neutrals that can be utilized for nanoscale surface modification of polymeric and textile substrates, keeping bulk property unaltered. Plasma modification with a non-polymerizing gas or precursor causes surface etching, activation, changes in surface energy, oxidation, cleaning and increase in surface roughness. Similarly, plasma reaction with a bigger unsaturated to saturated molecule leads to plasma polymerization, coating, chemical vapour deposition and creation of nanostructures. These above two plasma processes ultimately could improve the functional values of the textile substrates in terms of improvement in oil absorbency, oil repellency, UV protection, water absorption, water repellent, antimicrobial, FR, adhesion, anti-static, desizing, dyeing, printing and anti-felting of wool [34–38]. Non-thermal plasma can be considered for the above value-added modifications of textile and other polymeric substrates. Plasma is an energetic chemical environment, where the generation of plasma species opens up diverse chemical reactions resulting in various end applications and the current chapter summarizes the application of water-free plasma technology for FR and water repellent finishing of textiles.

3.1.1 Flame Retardant Finishing Using Plasma

Argon plasma has been utilized for multifunctional finishing of cotton textile to impart flame retardancy and water repellency functionalities in a single step [39]. Graft copolymerization of monomers of acrylate phosphate and phosphonate derivatives for FR effect along with carbon tetrafluoride (CF_4) for hydrophobic effect was carried out in the presence of argon plasma. In such combined finishing, fluorocarbon gas (CF_4) had no detrimental effect on fire retardant finishing. The same research group from the national textile centre has also reported the plasma pre-activation cotton surface followed by plasma polymerization of cyclotriphosphazene and hexa-chlorocyclotriphosphazene by passing the fabric through the plasma zone to accomplish FR nanolayer on the fabric surface. In addition, they have also reported the argon plasma induced graft copolymerization of monomers, like phosphorous diethyl phosphate and diethyl-2-(methacryloyloxyethyl) phosphate on poly (acrylonitrile) (PAN) fabric. Different metal and metal oxide mixtures, like nanosilver, titanium dioxide and aluminium oxide were applied in the argon plasma-pre-treated polyester fabric, and the fabrics could exhibit thermal stability [40]. Furthermore, the modified fabrics could also show good antimicrobial efficacy. Graft polymerization of 5 % phosphorous-containing phosphate and phosphoramidate monomers in argon plasma followed by sulphur hexafluoride (SF_6) plasma treatment could impart the excellent flame retardancy as well as water repellency in silk fabric, where the $-CF$, $-CF_2$ and $-CF_3$ molecules were mainly responsible for improvement in hydrophobicity [41]. Like above, i.e. application of metal or metal oxide particles, cotton fibres were coated with silicon dioxide (SiO_2) layers in the presence of atmospheric pressure plasma for its application in upholstery furniture, clothing and military. The SiO_2 network armour was obtained through hydrolysis and condensation of the precursor TEOS, and it was sequentially cross-linked on the fibre's surface; thus, due to the presence of SiO_2 network armour, thermal stability of the fabric was improved. The applied finish was found to be durable even after intense ultrasound washes. A thin film was deposited in the pre-mixture of oxygen gas and tetramethyldisiloxane (TDMS) monomer [42]. Such deposit can be used for FR modification of polyamides and polyamide nanocomposites. The LOI value of the coated polyamide-6 nanocomposites was profoundly enhanced, when the film thickness was equal to 0.6 mm, possibly due to formation of surface protective layer during burning. The very presence of carbonaceous as well as silica-like layers together acts as a barrier, thus restricting heat and mass transport from the burning zone. Similarly, low-pressure microwave plasma was utilized for graft polymerization of fluorinated acrylate monomer 1,1,2,2-tetrahydroperfluorodecyl acrylate (AC8) on polyamide 6 (PA6) to decrease rate of heat release by 50 %. This was happened owing to gas phase reaction of $CF•x$ radicals with the different fragments of polymer generated during the thermal degradation, resulting in dilution of the combustible gases [43]. Joëlle et al. reported the use of some monomer along with diethyl (acryloyloxyethyl) phosphoramidate (DEAEPN) monomer, ethyleneglycoldiacrylate (EGDA) and photo-initiator, for simultaneous dyeing and FR finishing of cotton textile [44]. A recent patent described the application of atmospheric plasma for flash-fire resistance finishing of cotton and poly (meta-aramid) fabrics using clay and hexamethylene disiloxane (HMDSO) in different combinations. The effect of presence of clay compound was clearly visible in this experiment [45]. Low-pressure plasma was used for polymerization of 1.1.3.3-tetramethyl disiloxane (TMDS) monomer in oxygen gas mixture to deposit a thin film on polyamide-6 (PA6) substrates [46]. The availability of oxygen during the polymerization process of TMDS stimulated the formation of more thermally stable coatings with efficient fire retardant attribute. Similarly, plasma treatment in the presence of phosphorus compound can be used for FR finishing of cotton, acrylic and viscose rayon fibres [47].

3.1.2 Hydrophobic Finishing of Textile Using Plasma

Water repellent or hydrophobic finishing of textiles is important in order to ensure unwanted staining, wetting or chemical contamination in the presence of liquids in the form of rainwater, chemicals, tea, coffee, food, beverages and pesticides. The water repellent functionality in textile does not allow the liquid droplet to get absorbed by the fabric or allows the water droplets to roll off from the surface. In the hydrophobic functionalization of textiles, the surface energy plays an important role, where surface energy of the base (textile) material is reduced by plasma coating, plasma deposition, chemical coating, lamination and layer-by-layer deposition using hydrocarbon, fluorocarbon or silicone molecules. Samanta et al. demonstrated application of helium and tetrafluoro ethane (TFE) plasma at atmospheric pressure for hydrophobic finishing of cellulosic substrates [48]. The water absorbency time was found to be higher than 60 min after 2 min of plasma treatment. Indeed, the sample could also exhibit super-hydrophobic characteristic as the water contact angle was higher than 150° with significantly lower water rolling angle of 5°. The imparted finished was found to be durable to several washing cycles. Atmospheric pressure cold plasma was also generated in the presence of fluorocarbon gas to make cotton textile hydrophobic. In this direction, plasma reaction of difluoromethane and pentafluoroethane could lead to development of water repellent cotton textiles, as shown in Fig. 4 [49]. After 1-min plasma treatment, water absorbency time increased to 2.5 min from 0.05 min (3 s) in the untreated sample. With increasing plasma treatment time, water absorbency time also increased linearly. The improvement in water absorbency time with plasma treatment signifies that more amount of fluorocarbon gas was responsible for better hydrophobic finishing of textile. In the 3-min plasma-treated sample, with carrier (He) to fluorocarbon (FC) precursor gas flow ratio of (3) with total fluorocarbon gas flow of 600 ml, the water absorbency time was 9 min. However, it was interesting

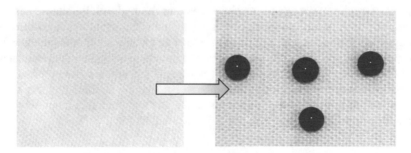

Fig. 4 Conversion of hydrophilic cotton into hydrophobic by plasma treatment [49]

to observe that when the carrier to precursor gas flow rate was kept at 12, while keeping the total fluorocarbon gas flow constant at 600 ml, the sample exhibited 40-min water absorbency time. Water contact angle value was also significantly increased from $\sim 0°$ in the untreated sample to 142° in the 12 min plasma-treated sample. After plasma reaction, hydrophilic cotton turned to hydrophobic possibly due to the attachment of different fluorocarbon species such as $-F$, $-CF$, $-CF_2$ and $-CF_3$ with cellulose polymer. Sulphur hexafluoride (SF_6) plasma could increase the hydrophobicity of PET, silk, cotton and cotton silk-blended woven textiles [50]. After plasma modification, the water contact angles were found to be 135°–145°, whereas water could easily absorbed and spread by the untreated fabric. The SF_6 cold plasma was also used in order to impart a high degree of water repellence in Bombyx mori silk fabric [51]. Another silk fabric, mulberry silk, was helium–fluorocarbon (He–FC) plasma treated at atmospheric pressure for 1 min and the water absorbency time increased from 2 min in the untreated sample to 14 min in the plasma-treated sample [52]. The CF_4 plasma could produce a Teflon-like structure with high degree of water repellence attributes on the polyethylene terephthalate surface with water contact angle of 120°–155°. Similarly, a barrier coating on high performance aramid fabrics was accomplished using hydrogen and hexafluorethane (C_2F_6) plasma [53].

Similar to above, i.e. application of fluorocarbon molecules for hydrophobic finishing of textile substrates, the hydrocarbon monomer, like methane (CH_4), ethylene (C_2H_4) or acetylene (C_2H_2), was also used to deposit the film-like surface coating of cross-linked amorphous hydrocarbon layers, and thus water repellence was imparted in textile substrates [54]. Plasma polymerization in the presence of CH_4 gas could deposit a very smooth hydrocarbon layer composed of CH_2 and CH_3 groups [55]. Atmospheric pressure cold plasma was generated in the mixture of 1,3-butadiene gaseous unsaturated monomer and helium (carrier gas) for water repellent finishing of the cellulosic textile [36]. A plasma reaction of 1.5 min resulted in hydrophobic conversion of the otherwise hydrophilic cellulosic substrates. As a result of this water absorbency time improved from less than a second in the untreated sample to about 28 min in the plasma-treated sample. A minimum 7-min plasma reaction was required to achieve water absorbency time of more than 60 min. The mechanism of plasma reaction was proposed based on the results of SIMS, GC-MS, OES and XPS analysis. Parida et al. [56] recently reported the reaction of cyclic molecule (styrene) in reducing the hydrophilic characteristic of cotton textiles. In the plasma-modified sample, the water drop disappearance time and the water contact angle improved to one hour and 133°, whereas these values were only 4–7 s and $\sim 0°$ in the untreated sample. When a bigger molecule, such as vinyl laurate, was used for hydrophobic finishing of cotton textiles, wash durable water repellent functionality was possible to achieve [57]. Similarly, the plasma reaction of dodecyl acrylate can be used for hydrophobic finishing of viscose textile, which is durable aqueous and solvent washing [58].

3.2 Application of UV Technology in Textile

Similar to plasma treatment of textile as ascribed above, UV treatment can also be used surface modification of natural as well as synthetic textiles without usage of water. Among the various water-free technologies for surface modification of textiles, UV irradiation has been found to be one of the important techniques, as the process is cost-effective, clean, eco-friendly and energy saving [31]. Among the different UV sources, excimer lamps can be used for surface modification of polymers and textile substrates in order to improve functionality like wettability, adhesion strength, antimicrobial, anti-static, anti-soiling, anti-felting, pilling resistance and cross-linking properties. Lasers that use a noble gas for its generation are generally referred to as excimer lasers; they operate in the UV (193 nm) to near UV region (351 nm). The bond dissociation power of the photon is different for the different excimer systems, e.g. if xenon gas is used, it could emit 172 nm light with photon energy of 7.2 eV, suitable for breaking of polymeric bonds. The surface of the wool protein fabric was modified using a xenon excimer UV lamp with a wavelength of 172 nm [59]. After the UV irradiation in air atmosphere, wetting time get reduced from 280 s in the untreated sample to 22.2 and 1.5 s in the 1-min and 5-min treated samples, respectively. Wetting time was found to decrease exponentially initially and, later on, linearly with UV increasing irradiation time. Similar trends were also observed, when the samples were treated in the presence of O_2 and N_2 gaseous. In a wicking experiment, the UV-modified wool fabric took only 10 s to travel a height of 2 cm irrespective of gases used in comparison with 2 min required by the untreated sample [60, 61]. Periyasamy et al. [62] reported the application of 172 nm UV excimer irradiation to improve the wetting and wicking properties of silk fabrics. The improvement in in-plane and vertical water transport was occurred possibly due to the increased in surface roughness and formation of nanopores in the treated sample. A multifunctional woollen fabric with hydrophobic functionality on one face and hydrophilic functionality on the opposite face were developed by first padding the textile with a mixture of NUVA HPU (fluoroalkyl acrylic copolymerisate emulsion) resin [31]. Thereafter, one side of the hydrophobic fabric was irradiated under 172 nm UV excimer lamp in order to develop hydrophilic characteristic that ultimately helps to develop smart dual functional hydrophilic–hydrophobic surfaces in a single piece of fabric. Water contact angle on the irradiated side, i.e. the hydrophilic side, decreased from 60° to 10° with an increase in the UV irradiation time by 30 min. It was interesting to note that the non-irradiated fabric surface could maintain hydrophobic characteristic with a contact angle of as high as 130°.

The UV as a pre-treatment has been explored for surface modification of protein fibre for its improvement in dyeing. After UV irradiation, the acid dye uptake was increased significantly [59]. After 20 min of dyeing, the dye exhaustion percentage increased to 89.1 % in the 1-min UV-irradiated sample in nitrogen atmosphere from 72.4 % in the non-irradiated sample. Similar to dye exhaustion, the K/S value was also found to follow a similar trend. The dye uptake in the treated samples was always higher in comparison with the untreated samples [63]. Samples irradiated in different gaseous environments exhibited higher dye uptake than their respective untreated samples. In another study, 253.7 nm UV irradiation was used for 40 and 60 min for surface modification of wool fabric and different properties were examined [64]. Similar to acid dyeing of wool, when the UV-irradiated samples were dyed with reactive dyes, the K/S value was also found to increase [63]. The UV irradiation might have helped in etching, chemical modification and formation of C–O, C=O, O–C=O, C–O–O and –OH groups that are responsible for the higher dye uptake [60, 65]. When nitrogen gas was used, there was additional formation of $-NH_2$ groups that played a favourable role in dyeing with anionic reactive dyes. On the other hand, a negative phenomenon was noticed, when the UV-irradiated sample in N_2 atmosphere was dyed with basic (cationic) methylene blue dye. Like plasma treatment of polymer and textile substrates, the graft copolymerization of methacrylamide on the UV-irradiated cotton textile was investigated using benzophenone as photosensitizer to improve the thermal stability of the fabric [66].

4 Sustainable UV Protective Finishing of Textile

A short UV exposure is considered to be appropriate, as it could improve the body's resistance to various pathogens, despite the fact that an excessive UV radiation exposure causes premature ageing, sagging, roughening, wrinkles, inflammation of human skin, blotches, erythema or sunburn and basal cell cancer [67]. Approximately, 6–7 % UV radiation present in the solar spectrum reaching to the earth causes several skin diseases [68]. Therefore, the UV protective finishing of textiles is an emerging area and getting a considerable attention so as to design requisite personal protective healthcare textiles. UV protection of a textile material is measured in term 'UV protection factor' (UPF: $Risk_{unprotected}/Risk_{protected}$) and it signifies that how long a person wearing the textile can stay in the sun before skin reddening starts, compared to a person without the said textile as a cover [67, 69, 70]. The UPF of a textile depends on the fibre used, depth of colour, fabric design, moisture content and the presence of other chemicals like optical brightening agents, finishing agents, UV absorber and laundering conditions [68]. A fabric with an UPF value of 40–50 or more is considered as excellent protection from harmful UV rays. In the past, various well-known chemicals, such as silicate, optical brightening agents, derivatives of o-hydroxyphenyl diphenyl triazine, triazine class-hindered amine, and the nanoparticles, like TiO_2 and ZnO, have been utilized to improve the UV protective performance of a textile substrate [71–74].

It was observed that a plain woven-bleached cotton fabric showed poor UPF value of 10, and after mordanting with tannic acid and alum, the UPF value remained unaltered. However, after application of BPS in alkaline condition, the UPF value was found to improve more than 100. This translates into the reduction in UVA and UVB transmittance percentages from 9.5 to 1.2 % and from 7.2 to 1 %

in the untreated to BPS-treated samples, respectively. If the bleached fabric is directly treated with alkaline BPS solution without any pre-mordanting, the same sample showed a lower degree of UPF value. Thus, the improvement in UV protection was arisen due to the synergistic of mordant and N, N-alkyl benzeneamine, present in BPS [29, 75]. The imparted finished was found to be durable up to two washing cycles. Similarly, UPF value of SJ, an alternative natural ingredient, when applied in cotton fabric, was also improved to more than 50; the UVA and UVB transmittance percentages were only 2 and 1.8 %, respectively, in the treated sample. In this sample, the UV stability was enhanced due to the presence of silicate molecules and organic colour compounds. Similarly, banana peel sap was attempted for improving the UV protective performance of cotton textile along with improvement in natural colour and antimicrobial efficacy, where the UPF value improved from 19.8 to 60 in the untreated to treated samples, respectively [76]. In the similar direction, other natural ingredients and naturals dyes, like madder wades, knotgrass, fenugreek, marigold, babool, manjistha, annatto, ratanjot, indigo, mulberry fruit extract, sweet spring citrus oil, honeysuckle extract, grapes oil, eucalyptus leaf, Flos Sophorae extract, have been explored for improving the UV-blocking properties of jute, cotton, bamboo, tencel fibre, cotton/polyester fibre blend, tencel/polyester fibre blend, silk, wool, linen, hemp and many more textile substrates [68, 77–83]. Similar to UV protective finishing of textiles using these plant molecules, some biomaterials, biomolecules, natural polymer and polymeric nanomaterials, like silk sericin, jojoba oil, aloe vera and nanolignin, were also attempted for UV protective finishing of polyester and other textile substrates [71, 84–86]. Besides their ensured UV protection, they could also serve as antimicrobial, anti-static, antioxidant and natural dye.

5 Textiles with Fragrance Finish

Nowadays, people frequently get stressed to accomplish many time bounded responsibilities. To keep such physical and mental stresses in manageable situation and also to live a peaceful and comfortable life, they normally prefer to be a part of yoga, spa, exercise and leisure. In this context, value-added traditional to functional home and apparel textiles play an important role by ensuring a mind-refreshing fragrance with a fresh and energetic feel to the wearer of such specially textiles. Apparel and home textiles, like bed linens, T-shirts, shirt, pillow covers, socks, inner garments and bed sheets, do not remain clean and fresh due to their everyday usage and frequent exposure to sweat, hot and humid environment that offer a favourable situation for microbial attack, resulting in production of foul smell. This is happened, because when such textiles are used, the body temperature is ideal for the growth of bacteria. Therefore, the aromatic and medicinal plant extracts, other natural ingredients and synthetic environment-friendly chemicals with amazing fragrance attributes are required to be incorporated to those textiles so as to ensure energetic and pleasant fresh feel the wearer, while masking foul smell. Such kind of speciality textile could also be used to decor our living and bed rooms providing a much needed relief to the otherwise our busy life. Aroma is an oil-based compound extracted from the plant products as well as synthetic materials. Application of essential ingredients to reduce the microbial attack on textile substrates has become an important aspect of textile finishing [87]. Fragrance finishing of textiles is also a similar process that enhances the value of the textile products by adding various soothing scent to it. In fragrance finishing of textile, microencapsulation process has been emerged as a useful method for protecting various aromatic or other functional ingredients on textile substrate by reducing evaporation or leaching. As the capsules do not have any affinity towards the fabrics, suitable binder may also be used to fix the capsules on textile structure to achieve better wash durability [88]. During the actual usage of such textiles, the microcapsules get busted either by pressure, friction or rubbing action and the fragrance molecules get diffused in the environment, thus making the local atmosphere pleasurable [89].

To fulfil the market demand of such high-valued textiles with assured aroma functionality, Bombay Dyeing has launched a new aroma finish bed linen made of soft cotton satin that comes with natural lavender, jasmine and rose water that revitalizes our senses, rejuvenates our soul, gives pleasant feeling and refreshes the ambience. In this direction, the beauty companies have begun to incorporate aromachology or aromatherapy into their product development. The US retail sales of all home fragrance products that are divided into environmental fragrance and fresheners climbed nearly 11 % to $1.96 billion in 1999, over in 1998 sales of $1.77 billion according to information from Kalorama, a New York City-based market research firm [90]. The investigation of US retail sales of home fragrance exhibits that sales increased from $1.35 billion in 1995 to $1.96 billion in 1999 and are projected to increase by 7.9 % to $2.87 billion in 2004. Some of the preferred aromas for textiles and home furnishing applications are rose, sandalwood, lavender, jasmine and champa that contain active ingredients like santalols, fusanol, santene, teresantol, benzyl acetate, linalool, linalyl acetate, benzyl benzoate and geraniol [79, 91]. These biomolecules are beneficial in revitalizing our immune, nervous and brain psychological system, skin nourishing, smoothening of facial lines and wrinkles, cell regeneration, and as an anti-depressive and antiseptic agent in concurrence with their mind-blowing fragrance [91, 92].

Recently, an innovative approach of durable aroma finishing on lignocellulosic jute textile has been developed to suppress the typical unpleasant kerosene oil like smell of jute batching oil that seriously affects the consumers' acceptance of jute products [89]. The study envisages application of jasmine aroma oil on jute–cotton-blended fabric by microencapsulation technique from melamine–formaldehyde. The average size of the capsules was 253 nm and the presence of aroma was evaluated by sensory method. Durability of the finishing was improved after incorporation of DMDHEU, polyacrylates and polyurethane binders. In another study, methanolic extract of geranium, *Pelargonium graveolens* L' Herit. Ex Air. leaves, was used to impart antimicrobial as well as aroma finishes to cotton textile [93]. The active antimicrobial as well as aroma ingredients were microencapsulated in gum acacia (capsule wall) and applied in cotton fabric by two different methods, namely coacervation spray drying and direct spray drying techniques. The second process leads to achievement of smaller size microencapsules of 5 and 12.5 μ in powder form and aqueous form, respectively. The fabric could retain 50–60 % of the imparted aroma even after 10 washing cycles compared to nil retention in the pure geranium extract sample. In addition to aroma finish, fabric could also exhibit 70–77 % antimicrobial efficacy against *Staphylococcus aureus* and 25–40 % against *Escherichia coli* even after 15 washing cycles, whereas these values were nil in pure geranium extract-treated samples. As aroma is volatile substance, achieving a durable aroma retention on the textile substrates has been a long time dream for the textile chemist [94]. Aroma after microencapsulation can stay on textile for longer duration, as capsules greatly reduce the aroma evaporation rate. In this direction, Li et al. reported action of three fixing agents, three kinds of thermal curing equipments and various curing conditions on wash durability of cotton fabric treated with lemon fragrance capsules (45 % solid in water). It was found that acrylic binder might form better networks than the DMDHEU and polyurethane binders. Theoretically, stronger curing condition could cause increased evaporation of the aroma from the capsules. Conventional hot air tenter and infrared curing methods could produce samples that could withstand 25 washing cycles than the microwave-cured sample. Aromatherapic textiles using fragrance with β-cyclodextrin inclusion compound as a host molecule have also been developed, as β-cyclodextrin is beneficial for human body. It is capable of forming inclusion compounds with requisite fragrance molecules that fit into the cone-shaped hydrophobic cavity, resulting profound decrease in fragrance release rate. The study describes the sedative effects for emotion and the pharmaceutical effects of essential oils [90]. The sensorial results depict that the perfume of the fabric is sensed for after 30 days. Some of essential oils with the sedative effects are lemon, rose, cardamom, clove, jasmine, camphor, chamomile, mint, lavender, vanilla, sandalwood, balm, pine and patchouli that can have influence on emotions related to anxiety, lament, stimulation, anger, allergy, tension, irritability and melancholy [90]. Banupriya and Maheshwari [87] examined the effects of fragrance-finished fabrics by comparing the herbal and conventional methods. It has been postulated that the Rosa damascena herbal fragrance-finished fabric is better than conventionally finished fabrics, and such fabrics were mainly used during summer seasons. The treated fabric was also found to be very much hygienic with less fungi and bacteria. Specos et al. [95] reported the development and testing of two types of microcapsules containing essential oils for application in cotton fabrics by padding and coating methods. Microcapsules were obtained by complex coacervation using gelatine and arabic gum or by encapsulation in yeast cells in order to increase the durability of fragrances in textiles. Fragrance release characteristic from the microcapsules was performed with the help of electronic nose that consists of 12 gas sensors with different selectivity. A lemon essential oil encapsulated in gelatine–arabic gum microcapsules increased the fragrance durability in cellulosic substrate as compared to non-encapsulated essential oil, which can withstand only single washing. The morphology, durability of the fragrance and laundering properties of the treated fabrics were also investigated. Conventionally, fixation of microencapsules is carried out by thermal process, where a fabric is cured at 130–170 °C temperature for 1–10 min to tightly fix capsules on the fabric [88]. However, during the curing process, the aroma inside capsules can be lost through quick evaporation and swelling to escape or break the capsule that ultimately reduces the performance efficacy of aroma-finished textile. To address such aroma loss in thermal process, an UV resin was developed to fix the capsules, as the resin can be cured under UV light at much lower temperatures within few seconds only. During UV curing, the amount of depletion of aroma is limited, which ultimately improved the product quality in terms of aroma durability [88]. Under suitable finishing and curing conditions, the wash durability of finished fabrics can be more than 50 washing cycles, which is better than the thermal curing (25 cycles).

6 Summary

Chemical processing of textile is essential to ensure requisite value addition to textile substrates. In dyeing and functional finishing of textiles, industry consumes significant quantum of water and chemicals. They also produce large amount of effluent that is finally discharged into the water stream. Some of the present day's chemicals and auxiliaries used in textile preparatory processing, dyeing and finishing are not eco-friendly, and thus air, water and soil get polluted. Indeed, in the recent years, due to increased awareness on human health and hygiene, global warming, environmental pollution, climate change and carbon footprint, the demand of natural fibre-based textiles dyed and finished with natural ingredients, like natural dyes for coloration, enzyme for bio-polishing, neem and aloe vera for antimicrobial finishing, plant molecules for UV protection, and biomacromolecules and plant extracts for FR finishing is getting considerable attention in academic research and industrial product development. In the last few years, attempts have also been made on sustainable dyeing and value-added finishing of textiles using various plant/herbal extracts, biomaterials, biopolymers and biomolecules. As FR textile has direct relation with human health and material safety, it remains as one of the important finishing operations. The DNA molecule, whey proteins, casein, CFP, BPS, SJ and GCSE can be used as novel approaches for improving thermal stability of the cellulosic, lignocellulosic and protein textiles as well as paper substrates. In the BPS-, SJ- and GCSE-treated FR textiles, the LOI values were found to be >27 compared to their respective untreated sample with LOI values of 18 for cellulosic cotton fabric and paper, 21 for lignocellulosic jute fabric and 25 for protein wool fabric. UV protective finishing of textile is also important so as to prevent skin ageing, sunburn, blotches, wrinkles, sun tanning and skin cancer. A number of natural ingredients, like BPS, SJ, grape and mulberry fruit extract, natural dyes, chitosan, tulsi, silk sericin, nanolignin, aloe vera, honey, almonds, cucumber and mint, have been applied in textiles mainly for UV protection, skin care and infection control. In order to ensure a pleasant feel to the ambience, the extracts of jasmine, lavender, champa, sandalwood, etc., have been incorporated in textiles, by direct and microencapsulation techniques. Water-free plasma and UV technologies have been used as a pre-treatment, post-treatment, in situ reaction or post-polymerization for surface activation, oxidation, etching, polymerization, coating or deposition in order to impart different value-added functionalities in textile, such as water and oil absorbency, water and oil repellency, flame retardancy, UV protection, anti-static, anti-felting, dyeing and printing. As various natural ingredients have been established to be used as natural dyes, UV absorbers, FRs and fragrance, they can be effectively utilized for sustainable textile finishing.

References

1. Wang X, Lu C, Chen C (2014) Effect of chicken-feather protein-based flame retardant on flame retarding performance of cotton fabric. J Appl Polym Sci 1–8. doi:10.1002/APP.40584
2. Basak S, Samanta KK, Saxena S, Chattopadhyay SK, Narkar E, Mahangade R (2015) Flame retardant cellulosic textile using banana pseudostem sap. Int J Clothing Sci Technol 27 (2):247–261
3. Alongi J, Carletto RA, Blasio AD, Carosio F, Bosco F, Malucelli G (2013) DNA: a novel, green, natural flame retardant and suppressant for cotton. J Mater Chem A 1:4779–4785
4. Bosco F, Casale A, Mollea C, Terlizzi ME, Gribaudo G, Alongi J, Malucelli G (2015) DNA coatings on cotton fabrics: effect of molecular size and pH on flame retardancy. Surf Coat Technol 272:86–95
5. Alongi J, Carletto RA, Bosco F, Carosio F, Blasio AD, Cuttica F, Antonucci V, Giordano M, Malucelli G (2014) Caseins and hydrophobins as novel green flame retardants for cotton fabrics. Polym Degrad Stab 99:111–117
6. Alongi J, Camino G, Malucelli G (2013) Heating rate effect on char yield from cotton, poly (ethylene terephthalate) and blend fabrics. Carbohydr Polym 92:1327–1334
7. Schindler WD, Hauser PJ (2004) Flame retardant finishes. In: Schindler WD, Hauser PJ (eds) Chemical finishing of Textiles. Woodhead Publishing Limited, New York
8. Horrocks AR (2011) Flame retardant challenges for textiles and fibres. Polym Degrad Stab 96 (3):377–392
9. Horrocks AR, Price D (eds) (2009) Advances in fire retardant materials. CRC Press LLC, Cambridge
10. http://books.google.co.in/books?isbn=9067640824
11. Seymour BR, Denamin RD (1987) History of polymeric materials. In: The Google books
12. Katovic D, Sandra BV, Sandra FG, Branka L, Dubravko B (2009) Flame retardancy of paper obtained with environmentally friendly agents. Fibres Text Eastern Eur 17(3):90–94
13. Banerjee SK, Day A, Ray PK (1985) Fire proofing jute. Text Res J 56:338–343
14. Carosio F, Di Blasio A, Alongi J, Malucelli G (2013) Green DNA-based flame retardant coatings assembled through layer by layer. Polymer 54:5148–5153
15. Horrocks AR, Price D (2001) Book chapter on intumescent flame retardants for textile substrates. In: Horrocks AR, Price D (eds) Fire retardant materials. Woodhead Publishing Limited, New York, pp 45–67
16. Basak S, Samanta KK, Chattopadhyay SK, Das S, Narkar R, Desouza C, Saikh AH (2014) Flame retardant and antimicrobial ligno-cellulosic fabric using sodium metasilicate nonahydrate. J Sci Ind Res 73:601–606
17. Basak S, Samanta KK, Chattopadhyay SK, Das S, Narkar R, Desouza C, Saikh AH (2014) Flame retardant and antimicrobial jute textile using sodium metasilicate nonahydrate. Pol J Chem Technol 16(2):106–113

18. Alongi J, Carletto RA, Balsio AD, Cuttica F, Carosio F, Bosco F, Malucelli G (2013) Intrinsic intumescent like flame retardant properties of DNA treated cotton fabrics. Carbohydr Polym 96(1):296–304
19. Hull TR, Kandola KB (2009) Fire retardancy of polymers, new strategies and mechanism. Royal Society of Chemistry, RSC Publications, pp 001–020. doi:10.1039/9781847559210)
20. Basak S, Samanta KK, Saxena S, Chattopadhyay SK, Narkar R, Mahangade R, Hadge GB (2015) Flame resistant cellulosic substrates using banana pseudostem sap. Pol J Chem Technol 17(1):123–133
21. Basak S, Samanta KK, Chattopadhyay SK, Narkar R (2015) Self-extinguishable lingo-cellulosic fabric using banana pseudostem sap. Curr Sci 108(3):372–383
22. Basak S, Saxena S, Chattopadhyay SK, Narkar R, Mahangade R (2015) Banana pseudostem sap: a waste plant resource for making thermally stable cellulosic substrate. J Ind Text. doi:10.1177/1528083715591580
23. Basak S, Samanta KK, Chattopadhyay SK, Narkar R (2015) Thermally stable cellulosic paper made using banana pseudostem sap, a wasted by-product. Cellulose 22(4):2767–2776
24. Basak S, Samanta KK, Chattopadhyay SK, Pandit P, Maiti S (2016) Green fire retardant finishing and combined dyeing of proteinous wool fabric. Colora Technol 132(2):135–143
25. Basak S, Patil PG, Shaikh AJ, Samanta KK (2016) Green coconut shell extract and boric acid: new formulation for making thermal stable cellulosic paper. J Chem Technol Biotechnol. doi:10.1002/jctb.4903
26. Basak S, Samanta KK, Chattopadhyay SK (2014) Fire retardant property of cotton fabric treated with herbal extract. J Text Inst 106(12):1338–1347
27. Bosco F, Carletto RA, Alongi J, Marmo L, Di-Blasio A, Malucelli G (2013) Thermal stability of flame resistance of cotton fabrics treated with whey proteins. Carbohydr Polym 94 (1):372–377
28. Carosio F, Blasio AD, Cuttica F, Alongi J, Malucelli G (2013) Polyester and polyester cotton blend fabrics have been treated with caseins. Ind Eng Chem Res 53(10):3917–3923
29. Katarzyna P, Prezewozna S (2009) Natural dyeing plants as a source of compounds protecting against UV irradiation. Herba Policia 55:56–59
30. Neog SR, Deka CD (2013) Salt substitute from banana plant (*Musa Balbiciana* Colla). J Chem Pharm Res 5(6):155–159
31. Basak S, Samanta KK, Chattopadhyay SK, Narkar R (2016) Development of dual hydrophilic/hydrophobic wool fabric by Q172 NM VUV irradiation. J Sci Ind Res 75:439–443
32. Samanta KK, Basak S, Chattopadhyay SK (2014) Book chapter on environment-friendly textile processing using plasma and UV treatment. In: Muthu SS (ed) Roadmap to sustainable textiles and clothing. Springer, Berlin, pp 161–201
33. Samanta KK, Jassal M, Agrawal AK (2010) Atmospheric pressure plasma polymerization of 1,3-butadiene for hydrophobic finishing of textile substrates. J Phys: Conf Ser 208:012098
34. Samanta KK, Jassal M, Agrawal AK (2009) Improvement in water and oil absorbency of textile substrate by atmospheric pressure cold plasma treatment. Surf Coat Technol 203:1336–1342
35. Samanta KK, Jassal M, Agrawal AK (2010) Antistatic effect of atmospheric pressure glow discharge cold plasma treatment on textile substrates. Fibers Polym 11(3):431–437
36. Samanta KK, Joshi AG, Jassal M, Agrawal AK (2012) Study of hydrophobic finishing of cellulosic substrate using He/1, 3-butadiene plasma at atmospheric pressure. Surf Coat Technol 213:65–76
37. Wakida T, Tokino S, Niu S et al (1993) Characterization of wool and polyethylene terephthalate fabrics and film treated with low temperature plasma under atmospheric pressure. Text Res J 63:433–438
38. Panda PK, Rastogi D, Jassal M (2012) Effect of atmospheric pressure helium plasma on felting and low temperature dyeing of wool. J Appl Polym Sci 124(5):4289–4297

39. Safaek T, Grutzmachor L (2007) Multifunctional surfaces using the plasma induced graft polymerisation (PIGP) process: flame and waterproof cotton textiles. Surf Coat Technol 201 (12):5789–5795

40. Raslan WM, Rashed US, EI-Sayad H, El-Halwagy AA (2011) Flame retardancy using plasma nano technology. Mater Sci Appl 2:1432–1442

41. Kamalangkla K, Hodak SK, Grutzmacher JL (2011) Multifuctional silk fabrics by means of the plasma induced graft polymerisation process. Surf Coat Technol 205:3755–3762

42. Bourbigot S, Duquesne S (2007) Fire retardant polymers: recent developments and opportunities. J Mater Chem 17:2283–2300

43. Errifai I, Jama C, Bras ML, Delobel R, Gengembre L, Mazzah A, Jaeger RD (2004) Elaboration of a fire retardant coating for polyamide-6 using cold plasma polymcrization of a fluorinated acrylate. Surf Coat Technol 180(8):297–301

44. Grützmacher JL, Tsafack MJ, Kamlangkla K, Prinz K (2012) Multifunctional coatings on fabrics by application of a low-pressure plasma process. In: 13th International conference on plasma surface engineering, Sept 10–14, Garmisch-Partenkirchen, Germany, pp 26–29

45. Li YC, Schulz J, Mannen S, Delhom C, Condon B, Chang S-C, Zammarano M, Grunlan JC (2010) Flame retardant behavior of polyelectrolyte-clay thin film assemblies on cotton fabric. ACS Nanotechnology 4(6):3325–3337

46. Quede A, Jama C, Supiot P, Bras ML, Delobel R, Dessaux O, Goudmand P (2002) Elaboration of fire retardant coatings on polyamide-6 using a cold plasma polymerization process. Surf Coat Technol 67(5):424–428

47. Shah JN, Shah SR (2013) Innovative plasma technology in textile processing: a step towards green environment. Res J Eng Sci 2(4):34–39

48. Samanta KK, Basak S, Chattopadhyay SK, Gayatri TN (2015) Book chapter on water-free plasma processing and finishing of apparel textiles. In: Muthu SS (ed) Hand book of sustainable apparel production. CRC Press, Taylor and Francis, pp 3–37

49. Samanta KK, Patil PG, Saxena S, Arputharaj A, Basak S, Gayatri TN (2015) Value added nano-finishing of cotton textile using water-free plasma technology. Cotton Res J 7(1):83–92

50. Supasai T, Hodak SK, Paosawatyanyong B (2007) Effect of SF_6 plasma treatment on hydrophobicity improvement of fabrics. Jurnal Fizik Malaysia 28(1&2):1–6

51. Nimmanpipug P, Lee VS, Janhom S, Suanput P, Boonyawan D, Tashino K (2008) Molecular functionalization of cold plasma treated Bombyx mori silk. Macromol Symp 264:107–112

52. Teli MD, Samanta KK, Pandit P, Basak S, Gayatri TN (2015) Hydrophobic silk fabric using atmospheric prcssure plasma. Int J Biores Sci 2(1):15–19

53. Hocker H (2002) Plasma treatment of textile fibers. Pure Appl Chem 74(3):423–427

54. Girard-Lauriault PL, Desjardins P, Unger WES, Lippitz A, Wertheimer MR (2008) Chemical characterisation of nitrogen-rich plasma-polymer films deposited in dielectric barrier discharges at atmospheric pressure. Plasma Process Polym 5(7):631–644

55. Kim JH, Liu G, Kim SH (2006) Deposition of stable hydrophobic coatings with in-line CH_4 atmospheric RF plasma. J Mater Chem 16:977–981

56. Parida D, Jassal M, Agarwal AK (2012) Functionalization of cotton by in-situ reaction of styrene in atmospheric pressure plasma zone. Plasma Chem Plasma Process 32:1259–1274

57. Abidi N, Hequet E (2005) Cotton fabric graft copolymerization using microwave plasma. II. Physical properties. J Appl Polym Sci 98(2):896–902

58. Panda PK, Jassal M, Agrawal AK (2013) Functionalization of cellulosic substrate using He/dodecyl acrylate plasma at atmospheric pressure. Surf Coat Technol 225:97–105

59. Basak S (2012) Process development of wool fabric by 172 nm VUV Excimer lamp. Lap Lambert Publishing, Germany

60. Samanta KK, Basak S, Chattopadhyay SK (2014) Book chapter on environment friendly textile processing using plasma and UV treatment. In: Muthu SS (ed) Roadmap to sustainable textiles and clothing: eco-friendly raw materials, technologies and processing methods. Springer Publication, Berlin, pp 161–201

61. Basak S, Gupta D (2012) Double hydrophilic/hydrophobic wool fabric by using 172 nm Excimer lamp. Text Excellent 23:25–29

62. Periyasamy S, Gulrajani ML, Gupta D (2007) Preparation of a multifunctional mulberry silk fabric having hydrophobic and hydrophilic surfaces using VUV Excimer lamp. Surf Coat Technol 201:7286–7291

63. Basak S, Gupta D (2010) Functionalisation of wool fabric using 172 nm VUV Excimer lamp. Text Excellent 29:45–49

64. Xin JH, Zhu R, Shen J (2002) Surface modification and low temperature dyeing properties of wool treated by UV radiation. Color Technol 118:169–174

65. Maclaren JA, Kirkpatrick A (1968) Partially oxidised disulphide groups in oxidised wool-reactivity with thiols. J Soc Dyers Colour 84:564–567

66. Kaur I, Sharma RM (2007) Development of flame retardant cotton fabric through grafting and post grafting reaction. Indian J Fibre Text Res 32:312–318

67. Saravanan D (2007) UV protection textile materials. AUTEX Res J 7(1):53–62

68. Samanta KK, Basak S, Chattopadhyay SK (2015) Book chapter on sustainable UV protective apparel textile. In: Muthu SS (ed) Hand Book of sustainable apparel production. CRC Press, Taylor and Francis, pp 113–137

69. Menter JM, Hatch KL (2003) Book chapter on clothing as solar radiation protection. In: Elsner P, Hatch K (eds) Textiles and the skin. Curr Probl Dermatol. Basel, Karger 31, pp 50–63

70. Kursun S, Ozcan G (2010) An investigation of UV protection of swimwear fabrics. Text Res J 1–8. doi:10.1177/0040517510369401

71. Zimniewska M, Batog J (2012) Book chapter on ultraviolet-blocking properties of natural fibres. In: Kozlowski R (ed) Handbook of natural fibres: volume 2: processing and application. Woodhead Publishing Limited with The Textile Institute, pp 141–167

72. Mao Z, Shi Q, Zhang L (2009) The formation and UV blocking property of needle-shaped ZnO nanorod on cotton fabric. Thin Solid Films 517:2681–2686

73. Yadav A, Prasad V, Kathe A, Raj S, Yadav D, Sundaramoorthy C, Vigneshwaran N (2006) Functional finishing in cotton fabrics using zinc oxide nanoparticles. Bull Mater Sci 29 (6):641–645

74. Samanta KK, Saxena S, Vigneshwaran N, Narkar R, Kawlekar S (2012) Cotton textile incorporating titanium dioxide nanoparticles and method to manufacture the same. Indian Patent, Patent number 3468/MUM

75. Sayed EI, Mohamad M, Selim OY, Ibrahim IZ, Maha M (2001) Identification and utilisation of banana plant juice and its pulping liquors as anticorrosive materials. J Sci Ind Res 60:738–747

76. Salah SM (2011) Antibacterial activity and ultraviolet (UV) protection property of some Egyptian cotton fabrics treated with aquous extraction from banana peel. Afr J Agric Res 6 (20):4746–4752

77. Sun SS, Tang RC (2011) Adsorption and UV protection properties of the extent form honeysuckle into wool. Ind Eng Chem Res J 50:4217–4219

78. Chattopadhyay SN, Pan NC (2013) Development of natural dyed jute fabric with improved colour yield and UV protection characteristics. J Text Inst 104(8):808–818

79. Samanta KK, Basak S, Chattopadhyay SK (2014) Book chapter on eco-friendly coloration and functionalization of textile using plant extracts. In: Muthu SS (ed) Roadmap to sustainable textiles and clothing. Springer Publication, Berlin, pp 263–287

80. Sarkar AK (2004) An evaluation of UV protection imparted by cotton fabrics dyed with natural colorants. BMC Dermatol 4(15):1–8

81. Subramoniyan G, Sundarmoorthy S, Andiappan M (2013) Ultraviolet protection property of mulberry fruit extract on cotton fabrics. Indian J Fibre Text Res 38:420–423

82. Vijayalakshmi D, Ramachandran T (2013) Application of natural oil on light weight denim garments and analysis of its multifunctional properties. Indian J Fibre Text Res 38:309–312

83. Mongkholrattanasit R, Punrattanasin M (2012) Properties of wool fabric dyed with eucalyptus, rutin, quercitin and tannin by padding technique. In: RMUTP conference: textile and fashion Bangkok, Thailand

84. Aramwit P, Siritientong T, Srichana T (2012) Potential applications of silk sericin, a natural protein from textile industry by-products. Waste Manage Res 30(3):217–224

85. Gupta D, Chaudhary H, Gupta C (2014) Sericin-based polyester textile for medical applications. J Text Inst 105(5):1–11
86. Kozlowski R, Zimniewska M, Batog J (2008) Cellulose fibre textiles containing nanolignins, a method of applying nanolignins onto textiles and the use of nanolignins in textile production. WO 2008140337 A1, Application number PCT/PL2007/000025
87. Banupriya J, Maheshwari V (2013) Effects of aroma finish by herbal and conventional methods on woven fabrics. Text Sci Eng 3(3):1–3
88. Li S, Boyter H Jr, Qian I (2005) UV curing for encapsulated aroma finish on cotton. J Text Inst 96(6):407–411
89. Biswas D, Chakrabarti SK, Saha SG, Chatterjee S (2015) Durable fragrance finishing of jute blended home-textiles by microencapsulated aroma oil. Fibres Polym 16(9):1882–1889
90. Wang CX, Chen ShL (2005) Aromachology and its application in the textile field. Fibres Text Eastern Eur 13(6/54):41–44
91. Samanta KK, Basak S, Chattopadhyay SK (2015) Book chapter on speciality chemical finishes for sustainable luxurious textiles. In: Muthu SS, Gardetti MA (eds) Handbook of sustainable luxury textiles and fashion. Springer Publication, Berlin, pp 145–184
92. Samanta KK, Basak S, Chattopadhyay SK (2016) Book chapter on potential of ligno-cellulosic and protein fibres in sustainable fashion. In: Muthu SS, Gardetti MA (eds) Sustainable fibres for fashion industry, vol 2. Springer Publication, Berlin, pp 61–110
93. Thilagavathi G, Kannaian T (2010) Combined antimicrobial and aroma finishing treatment for cotton, using micro encapsulated geranium (*Pelargonium graveolens* L' Herit. Ex Air.) leaves extract. Indian J Nat Prod Res 1(3):348–252
94. Li S, Lewis JE, Stewart NM, Qian L, Boyter H (2008) Effect of finishing methods on washing durability of microencapsulated aroma finishing. J Text Inst 99(2):177–183
95. Specos MMM, Escobar G, Marino P, Puggia C, Victoria M, Tesoriero D, Hermida L (2010) Aroma finishing of cotton fabrics by means of microencapsulation techniques. J Ind Text 40 (1):13–32

Printed in the United States
By Bookmasters